Sensors 2013

AA003258

Topical Conference at the 2013 AIChE Annual Meeting: Global Challenges for Engineering a Sustainable Future

San Francisco, California, USA
3-8 November 2013

ISBN: 978-1-63439-026-2

Printed from e-media with permission by:

Curran Associates, Inc.
57 Morehouse Lane
Red Hook, NY 12571

Some format issues inherent in the e-media version may also appear in this print version.

Copyright© (2013) by AIChE
All rights reserved.

Printed by Curran Associates, Inc. (2014)

For permission requests, please contact AIChE
at the address below.

AIChE
3 Park Avenue
New York, NY 10016-5991

Phone: (203) 702-7660
Fax: (203) 775-5177

www.aiche.org

Additional copies of this publication are available from:

Curran Associates, Inc.
57 Morehouse Lane
Red Hook, NY 12571 USA
Phone: 845-758-0400
Fax: 845-758-2634
Email: curran@proceedings.com
Web: www.proceedings.com

TABLE OF CONTENTS

(294a) Designing Metabolite Biosensors for Bioprocess Monitoring Using Synthetic Biology

Polizzi, K. M., Imperial College London
Constantinou, A., Imperial College London
Goers, L., Imperial College London
Originally presented on: 11/5/2013 12:30:00 - 12:48:00

Synthetic biology is an emerging discipline which seeks to use engineering principles to design or redesign biological systems. One of the clear application areas for this technology is the design of new biomanufacturing systems for the production of useful compounds including specialty chemicals and biopharmaceuticals.

Sensors are key to any industrial manufacturing process, where they are used to monitor important variables that have been linked to product yield and quality and are often linked to process control systems that can take corrective actions if these variables are not within defined limits. Synthetic biology systems offer an opportunity to deploy biological versions of these sensors-- gene circuits and/or proteins which perform the same functions of monitoring and correction. These can be included as part of the system design and offer advantages in terms of ithe speed at which they can respond and the diversity of actions which can be selected.

Biosensors can be designed to exploit any part of biology as the recognition event including transcription, translation, or post-translational signal generation. Each of these has different advantages and utilities. In this talk, I will focus on two types of biosensors: transcription-based biosensors which utilise inducible promoters as the signal transduction element and post-translational biosensors based on Forster Resonance Energy Transfer (FRET) which use protein conformational change as a signal output. We have successfully applied both of these for monitoring the concentration of important small molecule metabolites in the production of biopharmaceuticals in mammalian cell culture. The targets of our biosensors include carbon and nitrogen sources and waste products generated by metabolism. I will cover all aspects of sensor design and validation from selecting the appropriate biological recogntion element through biosensors construction and testing. Finally, I will show how biosensor signal can be used predict more complex outcomes such as biomass or protein yield.

(294b) Development of Biosensors to Identify New Chemicals Against Dengue Fever Vectors

Shakalli Tang, M., Ohio State University
Wood, D. W., the Ohio State University
Originally presented on: 11/5/2013 12:48:00 - 13:06:00

The development of a bacterial biosensor containing a nuclear hormone receptor (NHR) has proven to be an effective alternative to current compound screening strategies that rely on hormone-mimicking properties. NHRs comprise one of the largest classes of protein targets because they can bind to a wide variety of endocrine disruptors in addition to their native hormones. Our allosteric biosensors can detect compounds that target NHRs and distinguish between their positive or negative effects on the receptor's activity. When this design is extended from human to insect NHRs, an opportunity arises to screen compounds for their potential role as insecticides against infectious disease vectors such as *Aedes aegypti*, the dengue fever mosquito. This project seeks to construct a nuclear hormone binding sensor in *E. coli* to favor the identification of those compounds that disrupt the endocrine system of this target insect. By incorporating the ligand-binding domain of *Aedes aegypti's* NHR 96 into our *E. coli* biosensing plasmid, we are able to detect its response to CITCO, a chemical compound known to be the NHR96 native ligand. After a thorough optimization of parameters such as incubation time, temperature, and dilution, we seek to attain a reproducible growth phenotype assay for this novel biosensing system. Furthermore, the specificity of a particular *A. aegypti* NHR to a given ligand will be harnessed to evaluate the compound's effectiveness as an insecticide.

(294c) New Molecular Probes for mRNA Detection

Blackstock, D.,
Chen, W., University of Delaware
Originally presented on: 11/5/2013 13:06:00 - 13:24:00

Numerous viral infections and life threatening diseases are characterized by a specific mRNA expression pattern, making it a reliable target for diagnostics and drug therapies. While focusing on the fast acting and highly specific characteristics necessary when creating worthwhile diagnostics, we have created a new generation of molecular beacons suitable for detecting specific mRNA molecules. The design tools include molecular beacon-type oligonucleotides in combination with zinc finger-fluorescent protein fusions for fluorescence signaling upon target binding. Ultimately the design allows for simple construction, easy operation, and modular mRNA targeting capabilities.

(294d) Aptamer Binding of Beta Amyloid for Early Detection of Alzheimer's Disease

Chan, H., Rensselaer Polytechnic Institute
Sorci, M., Rensselaer Polytechnic Institute
Agris, P., University of Albany, State University of New York
Belfort, M., University at Albany
Belfort, G., Rensselaer Polytechnic Institute
Originally presented on: 11/5/2013 13:24:00 - 13:42:00

Alzheimer's Disease (AD) is projected to be a major medical burden by 2050 and early detection will hopefully reduce this burden through prevention and, yet to be discovered, treatment measures. The leading candidates for the neurotoxic species causing AD are beta amyloid 1-42 ($A\beta_{1-42}$) oligomers. Detection in the cerebrospinal fluid of the variation in concentration of these structures and those of the protein tau could potentially be used for the diagnosis of AD. Here, we develop an affinity biosensor for early detection of AD based on specific RNA aptamer-$A\beta_{1-42}$ binding interaction. Aptamers were chemically immobilized and oriented onto a gold-coated crystal for measurement in a Quartz Crystal Microbalance with Dissipation (QCM-D). Both mass uptake (frequency change, Δf) and aptamer film rigidity (dissipation change, ΔD) were measured as a function of time to monitor aptamer immobilization and $A\beta_{1-42}$ binding. Sensitivity, selectivity and robustness of the $A\beta_{1-42}$ oligomer sensor will be presented.

(294e) Development of Anti-Capsular Polysaccharide Aptamers for High-Throughput Screening and Serotyping

Cress, B., Rensselaer Polytechnic Institute
Linhardt, R. J., Rensselaer Polytechnic Institute
Koffas, M. A. G., Rensselaer Polytechnic Institute
Originally presented on: 11/5/2013 13:42:00 - 14:00:00

Bacterial pathogens are often encapsulated by a viscous layer of polysaccharide known as the capsule, which serves as the first layer of protection against a diverse array of environmental pressures. The mucous-like layer formed by the capsular polysaccharide retains water to prevent desiccation and can even hinder opsonization and subsequent phagocytosis by host immune systems. A unique class of bacteria biosynthesize capsules that are identical to polysaccharides also produced in mammalian tissues and are thus non-immunogenic, further contributing to evasion of host immune response. Pathogens bearing these capsular polysaccharides are implicated in such human diseases as urinary tract infections, strep throat, and bacterial meningitis. Coincidentally these capsular polysaccharides are metabolic engineeering targets as they can also serve as precursors in the production of medically relevant glycosaminoglycans, or highly acidic linear polysaccharides such as heparin and chondroitin sulfate. The utility of a rapid and selective assay quantifying and differentiating between capsular polysaccharides is twofold. Engineered strains could be screened in a high-throughput manner and selected for increased capsular polysaccharide production levels. Furthermore, a molecular screen selective for specific polysaccharides could be utilized in disease diagnosis and in strain serotyping. As antibodies cannot be generated against these non-immunogenic polysaccharides due to their natural presence in host animals, we have instead opted to select DNA aptamers from an oligomer library against capsular polysaccharides purified from three distinct bacterial sources. Heparosan from *Escherichia coli* K5, chondroitin from *E. coli* K4, and hyaluronan from *Pasteurella multocida* were purified, as confirmed by NMR, from fermentation supernatant using anion exchange chromatography. Linkage of capsular polysaccharides to amine-functionalized magnetic microparticles by reductive amination with sodium cyanoborohydride was verified by using a method recently developed in our lab. Specifically, bead-immobilized capsular polysaccharides were digested into disaccharides with their cognate polysaccharide lyases, followed by 2-aminoacridone (AMAC) labeling and quantification in a single, highly sensitive reverse phase ultra-performance liquid chromatography-mass spectrometry (RP-UPLC-MS) experiment. Molecular beacons based on selected aptamers will be further developed for fluorometric high-throughput screens to evaluate the capsular polysaccharide production capacities of metabolically engineered microbes.

(294f) Use of Fluorescent Protein Fusions to Optimize Membrane Protein Expression in Anaerobic Photoheterotrophic Rhodobacter

Escotet, M. S., Pennsylvania State University
Erbakan, M., Pennsylvania State University
Kumar, M., The Pennsylvania State University
Curtis, W. R., The Pennsylvania State University
Originally presented on: 11/5/2013 14:00:00 - 14:18:00

We describe the use of fluorescent protein fusions as a method for optimizing and accurately quantifying membrane protein expression in Rhodobacter. Membrane proteins (MP) are a difficult class of proteins to study; however, their medical and biotechnical importance makes them compelling to address the associated challenges. Difficulties of expressing membrane proteins have been reported in number of other protein expression systems such as *Escherichia coli*, yeast and mammalian cell culture. The anaerobic photoheterotroph, *Rhodobacter sphaeroides*, has the desirable characteristic of large quantities of intracellular membranes (typically absent in prokaryotes) as a result of the need to assemble the photosynthetic apparatus which facilitates ATP formation in light, but without oxygen formation. This growth mode can be implemented in anaerobic photobioreactors where biomass yields of more than 9.2 grams per liter dry weight (gDW/L) have been achieved.

The production of the homologous water-transport membrane protein Aquaporin Z (rAqpZ) from *R. sphaeroides* was utilized as a model system to develop the screening approach. Expression levels were quantified by monitoring sample fluorescence and Western Blot densitometry (WBD). Initially, monomeric Banana YFP (mBanana) fusion was chosen as a fluorescent tag because antibodies are commercially available and there is no overlap between the fluorescence spectrum of the protein and that of *R. sphaeroides*. To demonstrate the utility of such correlation for optimizing protein expression, temperature effects were studied as a simple optimization parameter. Ambient and optimal growth temperatures (25°C and 32°C, respectively) for *Rhodobacter sphaeroides* were studied under anaerobic photoheterotrophic conditions, required to induce protein expression from the *puc*-promoter used to drive MP expression. Correlations between off-line sampled culture fluorescence and WBD were developed through the use of pure protein standards to demonstrate the feasibility of using fluorescence measurements as the screening method for quantifying membrane protein levels inside growing cells.

A substantially improved signal-to-noise ratio could be achieved by diluting cell suspension in bovine serum albumin (BSA) to provide for refractive index matching and adjust the relative magnitude of light scattering to absorption. Similar trends were observed in MP production per cell at both temperatures, although the productivity was much higher at 32°C due to faster growth. This work identified several opportunities for improving this screening method towards the development of an online monitoring method to track the expression of MPs *in vivo*.

Although the mBanana fluorescent fusion served as an effective method for correlating membrane protein concentration, these measurements could only be taken offline. The requirement for offline measurement results because variants of GFP (green fluorescent protein) such as mBanana are obligate aerobic fluorescent proteins and require oxygen exposure to fluoresce properly. This presents a significant experimental constraint due to the required anaerobic growth conditions. Maturation involves folding in the presence of oxygen, during which time the fluorescence level per protein increases. Measurements of mBanana-MP fusion maturation within Rhodobacter showed that mBanana required more than 40 hours to mature (20 times its literature value). This long maturation time further precludes the use of this protein for rapid and online monitoring. Ongoing research seeks to overcome this limitation by using a new generation of facultative anaerobic fluorescent proteins based on riboflavin-binding activation. We will present the effectiveness of using pGlow-Pp1 and pGlow-Bs1 derivatives commercially available from EvoCatal as fusion tags that will hopefully allow correlations with *in vivo* measurements as well as online productivity monitoring.

(294g) The Impact of Dissociation Constant On the Detection Sensitivity of Polymerization-Based Signal Amplification Reactions

Kaastrup, K., Massachusetts Institute of Technology
Chan, L.,
Sikes, H. D., Massachusetts Institute of Technology
Originally presented on: 11/5/2013 14:18:00 - 14:36:00

Many studies have demonstrated free-radical polymerization as a simple and inexpensive signal amplification technique for the detection of molecular recognition events indicative of disease states. Of the various chemistries that have been implemented, photopolymerization has shown promise as a point-of-care technology for its ability to generate micron-scale hydrogels in as little as 35 seconds. Photopolymerization-based amplification requires the localization of photoinitiators at a surface as a function of molecular recognition events. Much of the previous work developing this technique has explored the case in which the signal amplification rather than the binding kinetics of the molecular recognition is limiting. We provide the first systematic study of how dissociation constant impacts assay sensitivity using proteins with affinities typical of those encountered in molecular diagnostic applications involving protein-protein binding. We use experimental results to validate a mass-action kinetic model that may be used to predict assay performance by serving as an alternative/supplement to the currently employed empirical approach to polymerization-based amplification assay development. We show that this model, given the surface probe density and protein binding affinity, is able to predict whether or not the minimal surface initiator density required for polymerization will be exceeded at a particular protein concentration. In addition, the polymerization-based amplification assays were performed on novel activated agarose surfaces, an inexpensive alternative to the SAM modified glass surfaces typically used.

(294h) Creation of a Firefly Luciferase Reporter Cassette for Use in PCR-Mediated Gene Deletion and Fusion in Saccharomyces Cerevisiae

Ainsworth, W. B., Louisiana State University
Benton, M. G., Louisiana State University
Originally presented on: 11/5/2013 14:36:00 - 14:54:00

When exposed to environmental perturbations, organisms increase production of proteins implicated in survival. Assessing the transcriptional level of one or more of these responding proteins provides valuable information regarding the cellular surroundings. Reporter systems developed for quantification of promoter response typically rely on enzymatic reactions (β-galatosidase and luciferase) or fluorescent molecules (green fluorescent protein - GFP), however deficits in implementation (such as protein extraction for β-galactosidase and the necessity of expensive high-throughput machinery for GFP) severely limit biosensor applications. The development of *in vivo* luciferase reporter systems has led to inexpensive, sensitive and accurate promoter assays without the variability reported between *in vitro* samplings, but current luciferase reporter systems are largely inflexible to modifications to the promoter of interest. PCR-mediated gene deletion/complementation techniques that overcome promoter inflexibility have been developed for fluorophores-based reporter systems and show promise for extension to luciferase reporters. In our study, we report the creation of a novel vector system, which introduces a cytosol-localized *Photinus pyralis* luciferase [*LUC*(−SKL)*] capable of one-step, *in vivo* measurements into a promoter–reporter system via PCR-based gene deletion/complementation. Fluorescent microscopy is utilized to confirm cytosolic localization of the luciferase construct under *HUG1* promoter. Further, we demonstrate accurate promoter dose–response through comparisons of luciferase assays to that of a similar *HUG1*Δ::*yEGFP1* promoter–reporter system. Our data indicate that, alongside a significant decrease in experimental time, the luciferase reporter system is shown to be highly reproducible and correlate well to comparable reporter constructs. Though we have applied this construct directly to DNA damage biosensing, use extends to any application where promoter response levels are desired

(347a) Characterizing Human Airway Transepithelial Ion Transport By Organic Electrochemical Transistor Array

Yao, C., The Hong Kong University of Science and Technology
Xie, C., Division of Life Science, The Hong Kong University of Science and Technology
Huang, P., The Hong Kong University of Science and Technology
Hsing, I. M., The Hong Kong University of Science and Technology
Originally presented on: 11/5/2013 15:15:00 - 15:31:00

Organic electrochemical transistor (OECT) offers attractive characteristics suitable for measuring ion flows at biological-electronic interfaces. Recent examples for OECTs include cell biosensors to monitor cells' detachment[1] and epithelial tissue' integrity[2]. In this work, we report, to our knowledge, the first study of monitoring dynamics of transepithelial ion transport in human airway epithelial cells on an integrated OECT array.

For a proof of concept study, an OECT array consisting of 32 micro-OECTs was fabricated on a glass substrate. Conducting polymer, poly(3,4-ethylenedioxythiophene): poly(styrene sulfonic acid) (PEDOT:PSS), with good biocompatibility was used as the active layer where a direct contact was made with the basal laminar layer of epithelial cells. Selective patterned gold layers were served as source, drain and interconnect. Human airway epithelial Calu-3 cells were directly cultured on the OECT array surface. These cells express abundant cystic fibrosis transmembrane conductance regulator (CFTR) chloride channels in the apical membrane. The activation of CFTR channels results in a transepithelial chloride secretion followed by sodium flow through tight junctions because of electrical attraction, thereby a sodium concentration decrease in the intercellular space at the basolateral side. By taking advantage of the cation-sensitive property of OECT[3], transepithelial ion flows induced by CFTR channel opening are then adequately reflected by a change of channel current or effective gate voltage of OECT.

Using this approach, this integrated array platform consisting of epithelial cells on OECT has shown a very sensitive dose-response relation upon the addition of forskolin, a classical CFTR agonist. Moreover, the OECT device shown in this study exhibits good transfer characteristics and could be operated at very low voltage. This OECT-based platform has the potential to study ion transport in other physiological processes and can be used as drug testing platform to evaluate drug's effectiveness on the ion channel related diseases.

References:

[1] P. Lin, F. Yan, J. Yu, H. L. Chan, M. Yang, *Adv. Mater.* **2010**, *22*, 3655.

[2] L. H. Jimison, S. A. Tria, D. Khodagholy, M. Gurfinkel, E. Lanzarini, A. Hama, G. G. Malliaras, R. M. Owens, *Adv. Mater.* **2012**, *24*, 5919.

[3] P. Lin, F. Yan, H. L. Chan, *ACS Appl. Mater. Iinterfaces.* **2010**, *2*, 1637.

(347j) DNA-Based Microrna Sensors

Ehrhardt, K., The University of Texas at Dallas
Li, Y., The University of Texas at Dallas
Zhang, M. Q., The University of Texas at Dallas
Bleris, L., University of Texas at Dallas
Originally presented on: 11/5/2013 15:31:00 - 15:47:00

Studies over the past decade have demonstrated that differential expression of microRNA (miRNA) plays an important role in the pathogenesis of various cancers. Studies have also shown that miRNA affects cancer cells' response to treatment. Here we designed and built DNA-based sensors to study endogenous miRNA in mammalian cells. We constructed both direct and indirect sensors which report miRNA concentrations by producing a fluorescent protein. High miRNA concentration results in low fluorescence signal for the direct sensors and high fluorescence signal for the indirect sensors. We optimized our sensors in transient transfection experiments in HEK293 and TET-ON cells using fluorescent microscopy and flow cytometry measurements. By titrating synthetic miRNAs and using sensors for endogenous miRNA (hsa-miR-16, hsa-miR-17, hsa-miR-10b, and hsa-miR-146a), we show that the sensors report with high precision the miRNA concentrations. We also report time-lapse fluorescent microscopy experiments using inducible synthetic miRNAs. These results show that our sensors can reliably track miRNA concentrations. Finally, we use time-lapse data to build a quantitative model of our sensors and their response to miRNA. Using this model we created a map that relates binding affinity vs. miRNA abundance vs. sensor output. In conclusion, we present DNA-based miRNA sensors which reliably report and track miRNA concentrations.

(347c) Toward Electrochemical Screening of Pseudomonas Aeruginosa Antibiotic Susceptibility

Webster, T. A., Northeastern University
Sismaet, H. J., Northeastern University
Goluch, E. D., Northeastern University
Originally presented on: 11/5/2013 15:47:00 - 16:03:00

The emergence of antibiotic resistant bacteria is driving the need for novel antibiotics and new means of efficiently screening the efficacy of potential drug targets. Current antibiotic screening relies on overnight culturing of bacteria after exposure to the antibiotic. The health of the resulting biofilm is then typically measured by the production of fluorescent proteins or staining. As an alternative approach, we propose to utilize a microfluidic platform as a small volume assay to quantify bacterial susceptibility. Further, we plan to detect the electrical response of secreted electro-active molecules by the bacteria as they are exposed to antibiotics. It is expected that if the antibiotic being studied negatively impacts the growth of the biofilm then the production and excretion of electro-active molecules will decrease.

Pseudomonas aeruginosa, a common cause of nosocomial infections, represents a serious threat to immune compromised patients due to its resistance to several antibiotics and is often used as a model organism in biofilm studies. *P. aeruginosa* is an effective opportunistic pathogen in part due to the blue-green electro-active molecule, pyocyanin, that it continuously excretes. Pyocyanin production is up regulated with the production of biofilm.

In this study, *P. aeruginosa* biofilms were grown in polydimethyl siloxane (PDMS) microfluidic channels that were irreversibly bonded to disposable three electrode cells (EDAQ). Varying concentrations of *P. aeruginosa* cells in trypticase soy broth were loaded into the channels and cultured at room temperature (approximately 25°C). Fresh trypticase soy broth was continually flowed through the channel (approximately 100 nL/min via a syringe pump) providing a constant supply of nutrients to the cells. The pyocyanin concentration in the microfluidic chamber was measured over the course of several days by scanning potentials from -0.5 to 0 V versus a silver/silver chloride reference electrode via square wave voltammetry at a frequency of 15 Hz and amplitude voltage of 0.05 volts. It has been shown that the only electro-active molecule excreted by *P. aeruginosa* in this potential window is pyocyanin. An increase in the current signal at the oxidation potential of pyocyanin was observed over time signifying the growth of *P. aeruginosa* within the channel. Since the production of pyocyanin is correlated with the growth of *P. aeruginosa* cultures, this experimental approach can be used to quantitatively screen whether drug targets effectively disrupt biofilm formation in *P. aeruginosa* cultures. The use of a microfluidic setup with electrochemical detection allows for faster detection in a more realistic microbial environment.

(347d) Temperature-Dependent Electrical Properties of Graphene Inkjet- Printed On Flexible Materials

Kong, D., Stevens Institute of Technology
Le, L.,
Originally presented on: 11/5/2013 16:03:00 - 16:19:00

Graphene electrode was fabricated by inkjet printing, as a new means of directly writing and micropatterning the electrode onto flexible polymeric materials.
Graphene oxide sheets were dispersed in water and subsequently reduced using an infrared heat lamp at a temperature of ~200 °C in 10 min. Spacing between adjacent ink droplets and the number of printing layers were used to tailor the electrode's electrical sheet resistance as low as 0.3 MΩ/□ and optical transparency as high as 86%. The graphene electrode was found to be stable under mechanical flexing and behave as a negative temperature coefficient (NTC) material, exhibiting rapid electrical resistance decrease with temperature increase. Temperature sensitivity of the graphene electrode was similar to that of conventional NTC materials, but with faster response time by an order of magnitude. This finding suggests the potential use of the inkjet-printed graphene electrode as a writable, very thin, mechanically flexible, and transparent temperature sensor.

(347e) Blood Detection Using Biological Modified CNTs

Cortes, P.,
Olszewski, A.,
Fagan, D.,

Originally presented on: 11/5/2013 16:19:00 - 16:35:00

The present work has studied the synthesis process of biological modified Multi-Wall Carbon Nanotubes-based threads for the detection of human blood. In this work, Human Serum Albumin (HSA) has been targeted as an indicator of blood. A short chain peptide, GAQGHTVEK (GK-1) has shown to bind specifically to serum albumin over bodily fluids and everyday liquid media such as water, juice, milk and coffee. Here, GK-1 has been covalently attached to the surface of carboxylated MWCNTs through a carbodiimide catalyzed amide linkage, and the binding of HSA to the synthesized MWCNT-GK-1 complex has been confirmed. Common synthetic threads have been coated by the biological modified MWCNTs through a dipping process, resulting in a semi-conductive bio-sensing thread. The CNT-GK1-thread has also been incorporated in an electrical circuit assisted by a Data Acquisition system, and distinct changes in resistance across the textile fabric were observed during its exposure to blood, bodily fluids and various liquid media. Indeed, the fabrication of a stable CNT-textile capable of detecting human blood over different liquid media represents a useful sensing fabric for the law enforcement and military sector.

(347f) Ultrawide-Range Electrochemical Biosensing Using Electrospun Carbon Nanofibers With High Density of States

Mao, X., Chemical Engineering, Massachusetts Institute of Technology
Yang, X.,
Rutledge, G. C., Massachusetts Institute of Technology
Hatton, T. A., Massachusetts Institute of Technology
Originally presented on: 11/5/2013 16:35:00 - 16:51:00

We show that electrochemical sensors based on continuous and interconnected electrospun carbon nanofibers with high density of states (DOS) can achieve controlled increase in their electroactive surface areas (ESAs), and thus a dramatic expansion of electrochemical sensing ranges while maintaining high sensitivity. These high-DOS electrospun carbon nanofibers (HD-ECNFs) with excellent electrocatalytic activities can be obtained through optimization of carbonization conditions. The ESAs of the HD-ECNF sensors can be easily manipulated by adjusting deposition time during electrospinning. Dopamine (DA), an important neurotransmitter, is used as the model analyte to evaluate the performance of the HD-ECNF sensor. Compared to previously-reported CNT- and graphene- based electrochemical sensors for DA detection, the HD-ECNF sensor developed here can achieve a remarkably wider detection range, spanning more than six orders of magnitude (from 2×10^{-1} to 7×10^{5} μM), while exhibiting one of the lowest detection limits (8×10^{-2} μM).

(347g) Analysis Of Exhaled Breath For Diagnosis Of Tuberculosis

Li, M., University of Louisville
Knipp, R.,
Nantz, M. H., University of Louisville
Higashi, R. M., University of Louisville
Graham, J. E.,
Fu, X. A., University of Lousiville
Originally presented on: 11/5/2013 16:51:00 - 17:07:00

Analysis of exhaled breath promises to be a powerful tool for noninvasive diagnosis. Some carbonyl compounds have been reported as metabolic markers of Tuberculosis (TB). However, so far there is no established protocol for the analysis of carbonyl compounds in exhaled breath. We previously reported a preconcentrator approach for capture of carbonyl compounds in exhaled breath. However, the preconcentrator approach requires a long time for exhaled breath flowing through the preconcentrator and it is difficult to deal with infectious exhaled breath samples of TB patients. Here we reports a simple sampling chip with reactive aminooxy coating for the capture of carbonyl compounds in exhaled breath. The reactive silicon chip has thousands of micropillars in a microfluidic channel. The reactive chips were fabricated on 4"-silicon wafers. Deep reactive ion etching (DRIE) was used to create the channels and micropillars with a height of 100 μm. The micropillar surfaces were coated with a reactive aminooxy compound for trapping trace carbonyl compounds by means of oximation reactions. The captured ketone- and aldehyde adducts were eluted by ethanol and were directly analyzed by gas chromatography-mass spectrometry (GC-MS) and Fourier transform-ion cyclotron resonance (FT-ICR) mass spectrometry (MS). The results indicate that this method of breath analysis can be used for diagnosis of TB patients.

(347h) Towards the Detection of Endocrine-Disrupting Compounds: Using Whispering Gallery Mode Optical Biosensors to Detect Estrogen Mimics

Yang, Y.,
Hunt, H. K., University of Missouri
Originally presented on: 11/5/2013 17:07:00 - 17:23:00

Endocrine-disrupting chemicals (EDCs) can disturb the development of both the endocrine system, and of the organs that respond to endocrine signals, in organisms indirectly exposed to EDCs during prenatal and/or early postnatal life. Large numbers of potential EDCs have been released into the environment due to industrial processing. Unfortunately, many of these potential EDCs are poorly understood in terms of *in vivo* toxicity, chronic exposure impact, and threshold levels. The first step towards protecting human and animal health is to develop new experimental platforms capable of detecting potential EDCs on the basis of their interaction with known endocrine receptors. Here, we propose to use highly sensitive and specific, label-free whispering gallery mode (WGM) optical biosensors to identify trace amounts of potential EDCs in contaminated source waters. As a proof of concept, we will evaluate the sensitivity of the WGM biosensors towards the detection of diethylstilbestrol (DES), phthalate and nonylphenol. The surface chemistry of these devices is demonstrated using XPS, SEM, and fluorescent and optical microscopy. The optical performance of the biosensors is characterized by microcavity analysis. This work represents one of the first examples of the use of these highly sensitive devices for environmental pathogens, and will improve the detection methods used to monitor EDCs.

(347i) Gas Phase Organophosphorus Detection Via Encapsulation of Enzyme Into Peptide-Nanotubes

McDaniel, E., The University of Toledo
Kim, D. S., University of Toledo
Goltz, M., AFIT-AFB
Originally presented on: 11/5/2013 17:23:00 - 17:39:00

Gas Phase Organophosphorus Detection via Encapsulation of Enzyme into Peptide-Nanotubes

*Eric B. McDaniel, Dong-Shik Kim, Mark N. Goltz**

Department of Chemical and Environmental Engineering, The University of Toledo, Toledo OH 43606

**Air Force Institute of Technology, Wright Patterson AFB OH*

Abstract

Organophoshporus (OP) compounds are commonly used for various aggressive functions such as pesticides as well as nerve agents. These compounds are being used in the environment for crops and domestic use, and it is imperative to detect and remove these compounds to assure safety to humans as well as the environment. In this research, the goal is to develop an easy-to-make biosensor to detect these OP pesticides. By doing such, the sensors are prepared with enzyme-encapsulated Peptide Nanotubes (PNTs) that is made from diphenylalanine. PNTs can be formed by self-assembly, and are easily prepared. From other research, PNTs have been found to help preserve the enzyme activity. Organophosphorus Hydrolase (OPH), an enzyme that has been used to detect Organophosphates, is used in this research as the enzyme to detect malathion. Malathion is a common OP pesticide that is mildly toxic as compared to its sibling OPs. Thus, using OPH gives a direct method of detecting malathion. These sensors are tested in liquid phase of malathion, using cyclovoltametry (CV) as the method of detection. The detection tests gives us the relationship that as the concentration of malathion increases, the increase of CV peak from the blank CV peak becomes larger. Thus this graph shows promising results that give a linear relationship of average percent increase of peak versus concentration. Enzyme activity tests of OPH that has been encapsulated into PNTs shows significant increase in longevity as compared to non-encapsulated OPH.

(374a) Simulation Tools for Nanoparticle-Based Composite Processing and Property Prediction

Bolintineanu, D. S., University of Minnesota
Lechman, J. B., Sandia National Laboratories
Schunk, P. R., Sandia National Laboratories
Originally presented on: 11/5/2013 15:15:00 - 15:40:00

We present a set of tools for the modeling of various aspects of solution-based nanomanufacturing processes. First, we discuss several discrete element modeling (DEM) techniques for the simulation of nanoparticle/colloidal suspensions, including efficient algorithms for the inclusion of hydrodynamic interactions, capillary forces, long-range colloidal interparticle forces, granular contact forces, solvent evaporation and highly aspherical particles. We apply these techniques to the simulation of a battery electrode manufacturing process, and show that the predicted microstructures provide relevant insights for the processing parameters. In the second half of the talk, we discuss methods for the prediction of the bulk properties of disordered, particle-based composites. We present several approaches that rely on a finite element model (FEM) representation of a sample of the material, which can be extracted from a DEM simulation or taken from a tomographic image of the real composite. Overall, this work presents an extensive, multiscale simulation toolset for connecting particle-based nanomanufacturing processing conditions to resulting microstructures and ultimately to material properties.

(374b) Tin Oxide Nanowires and Their Hybrid Architectures for Kinetically Fast Redox Couples in Dye-Sensitized Solar Cells

Vendra, V. K., University of Louisville
Nguyen, T., University of Houston
Druffel, T., Conn Center for Renewable Energy Research
Sunkara, M., University of Louisville
Amos, D. A., University of Louisville
Originally presented on: 11/5/2013 15:40:00 - 16:05:00

The redox electrolyte is an important component of the dye-sensitized solar cells. Iodide/tri-iodide electrolyte used in high efficiency cell is not suitable for practical application due to its corrosive nature, toxicity, competition between dye and tri-iodide for light aborption. One of the major drawback of using alternate redox couples is their fast recombination kinetics and mass transport limitations.

In this study, we show that tin oxide nanowires show a substantial improvement in the photovoltaic performance when compared to titania nanoparticles, when electrolytes having fast recombination kinetics are employed in dye-sensitized solar cells (DSCs). State of the art methods in literature employ titania nanoparticles use passivating layers of Al_2O_3 or poly(methyl siloxane) to lower the electron recombination rate. There have been reports investigating the performance of nanowire in conjunction with alternate redox couples. The single crystalline nature allows for fast electron transport in tin oxide nanowires. Further, prior work from our group has also shown that tin oxide nanowires exhibit two orders of magnitude higher electron lifetimes when compared to titania nanoparticles when iodide/tri-iodide is used as an electrolyte. In this study we employed tin oxide nanowires as photoanodes for DSCs and investigated their performance when kinetically fast electrolytes such as ferrocene/ferrocenium,tempo/tempo,ferricyanide /ferrocyanide are used as redox electrolytes.

Surface modifications such as $TiCl_4$ treatment and atomic layer deposition of titania and alumina were performed to further lower electron recombination in tin oxide nanowires. Fundamental properties such as electron transport time scales and electron recombination time scales were determined for tin oxide nanowires and hybrid architectures where the surface of tin oxide nanowires is coated with titania and alumina.

(374c) Multiscale Sustainability Assessment in Nanocoating Material Design and Manufacturing

Song, H., Wayne State University
Uttarwar, R., Wayne State University
Huang, Y., Wayne State University
Originally presented on: 11/5/2013 16:05:00 - 16:30:00

Nanopaint has drawn great attention in the coatings industry, as it may offer surface coatings with significantly improved performance or new functionalities and properties, such as surface texture alteration control, self-healing, toxic chemical/acid/corrosive agent repelling, glass opaqueness control, building exterior's infrared reflecting/absorbing properties control, and even pathogen growth control. Nevertheless, it is under scrutiny for the environmental and health risks which may pose in nanopaint manufacturing and nanocoating realization. Therefore, it is crucial to analyze the sustainability performance of nanocoating materials and prepare a roadmap for their successful development and application in the future.

In this paper, we introduce a systematic sustainability assessment methodology by resorting to multiscale science and sustainability science. Different from known methodologies focusing on the assessment of manufacturing process and manufactured product, this methodology initiates the assessment starting at the nanomaterial formulation design stage. A variety of material design parameters, such as those about polymer system, nanomaterial type, shape, size, and orientation, solvent and additives, are analyzed using a number of sustainability indicators, which will be incorporated into the analysis of coating manufacturing stages. Multiscale models are developed to predict nanoparticle emission severity in nanopaint spray and curing, nanocoating properties, product realization process performance on quality assurance, waste reduction, and operational safety. The efficacy of the multiscale sustainability assessment methodology will be demonstrated by the case studies on nanocoatings with super-scratch resistance performance and barrier property. The introduced methodology is generic and can be applied to study the sustainability performance of a wide range of nanocoating materials in their early design stage.

(374d) Continuous Nanoparticle Sizing and Characterization Via Microfluidic Interfacial Fluorescent Complexation

Meng, F., Texas A&M University
Ugaz, V. M., Texas A&M University
Originally presented on: 11/5/2013 16:30:00 - 16:55:00

A variety of characterization tools are available to sensitively analyze collected nanoparticles (e.g., condensation particle counters, differential mobility analyzers, scanning mobility particle sizers (SMPS), ion chromatography, mass spectrometry, and electron microscopy). But these methods often require a dedicated laboratory infrastructure to operate optimally, and are therefore not straightforward to adapt for online use. We have developed a new microfluidic-based approach that overcomes many of these limitations by harnessing the inherently steep chemical gradient established at the interface between co-flowing streams containing a nanoparticle-laden suspension and a fluorescent dye. This sharp mismatch acts to localize adsorptive dye-nanoparticle complexation interactions within a narrow interfacial zone, instantaneously producing an intense and easily detectable fluorescence signature. These interactions are inherently dominated by phenomena occurring at the nanoparticle surface, introducing the exciting possibility to extract information about particle size and morphology from the fluorescence profile. We demonstrate this capability experimentally using ZnO and TiO_2 nanoparticles, and introduce a reaction model that enables interfacial fluorescence to be predicted so that kinetic parameters associated with the underlying surface complexation can be quantified. Our approach offers potential for continuous operation that is particularly attractive in manufacturing settings where robust on-line characterization tools are needed.

(374e) Safer Formulation Concept for Flame-Generated Engineered Nanomaterials

Sotiriou, G. A., ETH Zurich
Gass, S.,
Cohen, J., Harvard University, School of Public Health
Pyrgiotakis, G., University of Florida
Pratsinis, S. E., ETH Zurich
Demokritou, P., Harvard School of Public Health
Originally presented on: 11/5/2013 16:55:00 - 17:20:00

The likely success or failure of the nanotechnology industry depends on the environmental health and safety of engineered nanomaterials (ENMs). While efforts toward engineering safer ENMs are sparse, such efforts are considered crucial to the sustainability of the nanotech industry. A promising approach in this regard is to coat potentially toxic nanomaterials with a biologically inert layer of amorphous SiO_2. Core-shell particles exhibit the surface properties of their amorphous SiO_2 shell while maintaining specific functional properties of their core material. A major challenge in the development of functional core-shell particles is the design of scalable high-yield processes that can meet large-scale industrial demand. Here, we present a safer formulation concept for flame-generated ENMs based on a one-step, in flight SiO_2 encapsulation process, which was recently introduced by the authors as a means for a scalable manufacturing of SiO_2 coated ENMs. Firstly, the versatility of the SiO_2-coating process is demonstrated by applying it to four ENMs (CeO_2, ZnO, Fe_2O_3, Ag) marked by their prevalence in consumer products as well as their range in toxicity. The ENM-dependent coating fundamentals are assessed and process parameters are optimized for each ENM investigated. The effects of the SiO_2-coating on core material structure, composition and morphology, as well as the coating efficiency on each nanostructured material, are evaluated using state-of-the-art analytical methods (XRD, N_2 adsorption, TEM, XPS, isopropanol chemisorption). Finally, the biological interactions of SiO_2-coated vs. uncoated ENMs are evaluated using cellular and animal bioassays, providing valuable evidence for reduced toxicity for the SiO_2-coated ENMs. Results indicate that the proposed 'safer by design' concept bears great promise for scaled-up application in industry in order to reduce the toxicological profile of ENMs for certain applications.

References

[1] Sotiriou G.A., Sannomiya T., Teleki A., Krumeich F., Vörös J. and Pratsinis S.E. Non-toxic dry-coated nanosilver for plasmonic biosensors. *Adv. Funct. Mater.* **20**, 4250-4257, (2010).

[2] Sotiriou G.A., Hirt A.M., Lozach P.Y., Teleki A., Krumeich,F. and Pratsinis S.E. Hybrid, silica-coated, Janus-like plasmonic-magnetic nanoparticles. *Chem. Mater.* **23**, 1985-1992, (2011).

[3] Gass S., Cohen J.M., Pyrgiotakis G., Sotiriou G.A., Pratsinis S.E. and Demokritou P. Safer Formulation Concept for Flame-Generated Engineered Nanomaterials. *ACS Sustainable Chem. Eng.* **in press**, DOI: 10.1021/sc300152f, (2013).

[4] Demokritou P., Gass S., Pyrgiotakis G., Cohen J.M., Goldsmith W., McKinney W., Frazer D, Ma J, Schwegler-Berry D, Brain J, Castranova V., An in vivo and in vitro toxicological characterisation of realistic nanoscale CeO_2 inhalation exposures *Nanotoxicology* **in press**, DOI: 10.3109/17435390.2012.739665 (2013).

(374f) Green Pathways for Development of Nanostructured Aerogel Photocatalysts Effective At Both UV and Visible Range

Li, H., University of South Florida
Sunol, S. G., University of South Florida
Sunol, A. K., University of South Florida
Originally presented on: 11/5/2013 17:20:00 - 17:45:00

Nano-structured highly porous $TiO_2/WO_3/Fe^{3+}$ aerogel photocatalyst are prepared, characterized, and tested for model photo-catalytic reactions. The catalyst structure is tailored to capture environmental pollutants and destruct them under both u-v and visible light through oxidation.

A novel and green method is utilized to prepare the unique surfactant templated aerogel photo-catalyst that has highly porous nanostructure with high surface area and tailored pore size distribution. Sol-gel process is combined with supercritical extraction, deposition and drying. Supercritical drying with heat treatment results in titanium dioxide anatase crystal form. Templates used further enable retention and tuning of the nano-pore structure and surface properties.

The catalyst was characterized using AFM, SEM, FIB, XRD, XPS, BET and porosimetry prior to using in reactions. The band gap of the catalyst particles were determined using diffuse reflectance.

Resulting aerogel $TiO_2/WO_3/Fe^{3+}$ has similar photo-catalytic capability compared to highly optimized commercial Degussa P25 under UV exposure and offer much superior photo-catalytic capability under visible light exposure. The model reaction utilized employed Methylene Blue photo oxidation under Visible and UV light.

(413a) Automatic Time-Dependent Flow Cut-Off System for Paper-Based Microfluidic Devices Using Erodible Polymeric Bridge

Jahanshahi-Anbuhi, S., McMaster University
Henry, A., McMaster University
Leung, V., McMaster University
Sicard, C., McMaster University
Pelton, R., McMaster University
Brennan, J. D., McMaster University
Filipe, C. D., McMaster University
Originally presented on: 11/6/2013 8:30:00 - 8:48:00

Abstract. An automatic time dependent flow cut-off system for volume control and reagent delivery applications in paper-based microfluidic devices has been developed. This system provides the ability to stop the fluid flow in porous media without external actuation through the use of a rapid dissolvable polymer film (referred to as a polymeric bridge). In this study, we quantitatively describe how this system responds to the dimensions of the polymeric bridge. Fluid crossing the polymeric bridge is stopped at distance that can be precisely pre-specified through the control of the dimensions of the polymeric bridge. In this work, we also show that the film can also be used as a reagent delivery device. This technology could be an excellent low-cost method for reagent delivery and volume control in paper-based analytical devices. Moreover, this method can be especially useful for time-sensitive or multi-step reactions/assays without compromising the simple, inexpensive, portable, and easy to use characteristics of paper-based microfluidic devices.

(413b) Enzyme Pegylation Photo-Patternable Hydrogels for Long-Term Implantable Glucose Sensors
Li, Z., University of Connecticut
Originally presented on: 11/6/2013 8:48:00 - 9:06:00

Enzyme PEGylation Photo-patternable Hydrogels for Long-term Implantable Glucose Sensors

Zhe Li[1], MS; Liangliang Qiang[1], PhD; Sagar Vaddiraju[1,2], PhD;

Fotios Papadimitrakopoulos[1,3], PhD[*]

1 Nanomaterials Optoelectronics Laboratory, Polymer Program, Institute of Materials Science, University of Connecticut, Storrs, CT 06269, United States

2 Biorasis Inc., 23 Fellen Road, Storrs, CT 06268, United States

3 Department of Chemistry, University of Connecticut, Storrs, CT 06269, United States

* E-mail address: papadim@mail.ims.uconn.edu

Abstract:

The function and lifetime of an implantable CGM device are intimately linked with the stability of the glucose oxidase (GO_x) enzyme, responsible for glucose detection. Enzyme denaturation, loss due to imperfect immobilization and biofouling-induced pore-clogging gradually decrease device stability. Poly(ethylene glycol (PEG)-ylated hydrogels offer an opportune venue to minimize biofouling, while retaining their highly hydrated state to prevent enzyme denaturation. These hydrogels, not only retain enzyme activity and provide good antifouling properties but also afford ease of fabrication *via* traditional photo-lithography methods.

A random copolymer based on PEGylated side chains together with cinnamyl ethyl methacrylate (CEMA) and glycidyl methacrylate (GMA) (namely poly(PEGMEM-CEMA-GMA)) was synthesized by free radical polymerization using azobisisobutyronitrile (AIBN) initiator. The polymer was mixed with GO_x enzyme and drop-casted on the electrode, followed by UV exposure for crosslinking and 24 hour incubation for reaction between the GO_x-amine and hydrogel's GMA-epoxide group. After enzyme PEGylation, GO_x

showed 1.7 times higher activity at 37°C, and when testing at 45°C, PEGylated GO_x preserved 2 times higher activity comparing with GO_x only.

Glucose sensors utilizing GO_x-grafted, poly(PEGMEM-CEMA-GMA) hydrogels showed excellent stability during continuous testing for 30 days. Control experiments with a hydrogel lacking the epoxide functionalities indicated that the majority of enzyme leached out within hours. By retaining >65% mole ratio in PEG functionality, these hydrogels provide a good biofouling resistance against serum albumin. The sensors exhibited a sensitivity of 320 nA $mM^{-1}mm^{-2}$ with an apparent *Michealis-Menten* constant of 25 mM of glucose and limit of detection of 1 μM. This hydrogel produces CGM sensors with long-term stability and optimal performance.

(413c) Performance of Implantable Glucose Sensors Coated With Dexamethasone-Releasing PLGA Microsphere/PVA Hydrogel Composites

Vaddiraju, S., Biorasis, Inc.
Wang, Y.,
Burgess, D.,
Papadimitrakopoulos, F., University of Connecticut
Li, Z., University of Connecticut
Originally presented on: 11/6/2013 9:06:00 - 9:24:00

Introduction: Implantable biosensors for continuous glucose monitoring hold promise for Diabetes care and management. The long-term potential of implantable biosensors is often negated by the foreign body response (biofouling and inflammation), elicited following device implantation. Herein, we show that outer sensor coatings based on dexamethasone-loaded PLGA microsphere/PVA hydrogel composite improves in vivo sensor performance by *(i)* suppressing inflammation and fibrosis *via* sustained delivery of dexamethasone following microsphere degradation; *(ii)* offsetting biofouling-induced permeability reduction *via* creation of macroscopic porosity following microsphere degradation.

Materials and Methods: Poly (lactic-co-glycolic) acid (PLGA) microspheres were prepared using a solvent evaporation technique. PVA hydrogels with and without PLGA microspheres were fabricated using a freeze/thaw cycling method. Implantable glucose sensors utilizing enzymatic detection of glucose were coated with dexamethasone -containing PLGA microspheres/poly (vinyl alcohol) (PVA) hydrogel composite. At first, the effect of the coating on the sensor linearity and sensitivity was determined *in vitro* (in PBS). Subsequently, coated and uncoated sensors (controls) were implanted into anesthetized rats using a thin wall needle (16 Gauge) and their sensitivity was determined periodically. All *In vivo* experiments were performed in accordance with the IACUC (Institutional Animal Care and Use Committee) guidelines.

Results and Discussion: *In vitro* results have indicated that the composite coating does not hinder sensor linearity but reduced sensitivity by 30%. *In vivo* studies have indicated that the uncoated biosensors suffered a rapid sensitivity loss in the first week post implantation and eventually failed to track the glycemic events. On the other hand, the coated sensors reproducibly tracked blood glycemic events even after 14 days following implantation due to the dexamethasone-induced inflammation suppression. Moreover, the sensitivity of the coated sensors increased with *in vivo* residence time which is currently attributed to macroscopic porosity generation following microsphere degradation (consistent with our *ex vivo* studies in porcine serum).

Conclusions: The PLGA microsphere/PVA hydrogel composite coatings did not compromise the linearity of the glucose biosensors, albeit formed a additional barrier to analyte diffusion thereby decreasing sensor sensitivity by 30%. *In vivo* studies indicated that the composite coatings were able to enhance sensor performance and lifetime. This renders these drug-delivering coatings as a promising vehicle for improving *in vivo* functionality of implantable biosensors.

Acknowledgements: Financial support for this study was obtained from US Army Medical Research Grants (W81XWH-09-1-0711 and W81XWH-07-10688), NIH grants (1-R21-HL090458-01, R43EB011886 and 9R01EB014586) and NSF/SBIR grants (1046902 and 1230148)

(413d) A Portable Ultra-Fast Response Gas Detector for Breath-By-Breath Measurement of Carbon Dioxide

Zhao, D., Arizona state university
Xian, X., Arizona State University
Miller, D., Arizona state university
Tsow, F., Arizona state university
Tao, N., Arizona state university
Forzani, E., Arizona state university
Originally presented on: 11/6/2013 9:24:00 - 9:42:00

The analysis of breath CO_2 provides valuable information of pulmonary and cardiovascular functions, and plays a crucial role in monitoring patients with respiratory problems. Most of the existing breath CO_2 monitors not only are expensive and bulky, but also require substantial training for correct operation and maintenance. In this work, we developed a low cost and high performance breath CO_2 analyzer for the personal use at home. This portable analyzer features an accurate colorimetric CO_2 sensor that can analyze end tidal CO_2 concentration in real time and a pressure transducer that can measure the tidal volume and breath frequency. The reversible colorimetric sensor changes color immediately (response time < 200 ms) from blue to yellow once exposed to expired air. The concentration of breath CO_2 can be determined by the color change of sensor cartridge. The pressure transducer measures expiratory flow rate, which can be used to determine the tidal volume and respiratory frequency through an intelligent signal-processing algorithm. The device was validated using a commercial capnometer based on infrared techniques and a good correlation was observed between the commercial CO_2 analyzer and our sensor. In addition, the device can be wirelessly paired to the users' cell-phone, computer, notebook, netbook or tabloid, providing a seamless information gateway between patients and medical care professionals and maximizing its benefit to the large and growing chronic respiratory disease patients.

(413e) The Use of Phage-Selected Peptides On Microcantilevers for Rapid Detection of Salmonella

Wang, J., Rice University
Karoonuthaisiri, N., National Center for Genetic Engineering and Biotechnology
Segatori, L., Rice University
Biswal, S. L., Rice University
Originally presented on: 11/6/2013 9:42:00 - 10:00:00

We report the use of a microcantilever sensor designed to measure the specific binding of *Salmonella* to immobilized phage selected peptide. An anti-*Salmonella* antibody is compared for specificity and selectivity against four peptides obtained from a phage display library. Particularly, the specificity of this sensor was investigated by distinguishing *Salmonella* from other bacteria (*Listeria monocytogenes* and *Escherichia coli*). We are also able to distinguish among eight *Salmonella* variants. A Langmuir isotherm model was applied to describe the molecular interactions on sensor surface and to assess the binding affinity constants of peptides. Both experimental and modeling results illustrate a peptide (MSal 020417) with better performance than commercially available antibody. In addition, various functionalization of different cantilevers on the same sensor chip allows simultaneous testing of multiple binding peptides and synchronous detection of multiple pathogens, which highly improves the efficiency of the detection. In summary, the microcantilever based technique provides a way to rapidly and accurately detect pathogens, as well as a high potential to be further developed as a screening method to identify pathogen specific binding candidates.

(413f) Detection of Cancer Biomarkers Using Hydrogel Microparticles

Al Ameen, M. A.,
Ghosh, G., University of Wisconsin-Madison
Originally presented on: 11/6/2013 10:00:00 - 10:18:00

Recent years have seen growing emphasis on early detection and identification of cancer risk in asymptomatic population to improve the clinical management of cancer. Despite the advances made in cancer therapy, the clinical outcome for cancer patients remains very poor. Early screening of cancer can facilitate in classifying asymptomatic people as likely or unlikely to harbor a tumor. Cancer molecular markers are assuming important role in cancer management including screening and early detection. Although ELISA remains the corner stone for protein quantification, adapting it for multiplexing or high through-put analysis has been difficult. By integrating antibody conjugation chemistry and immunoassay approach, we report the development of porosity tuned hydrogel (poly-ethylene glycol diacrylate (PEGDA)) microparticles for sensitive detection of molecular biomarkers which can be utilized for screening and early detection of cancer. We will also demonstrate that induction of porosity using PEG porogen improves the sensitivity of this simple hydrogel microsphere based system indicating that the sensitivity of the assay exceeds that of the conventional technologies.

(413g) High Sensitivity Dual-Analyte Biosensor for Trauma Patient Monitoring and Management

Kotanen, C. N., Clemson University
Guiseppi-Elie, A., Clemson University
Originally presented on: 11/6/2013 10:18:00 - 10:36:00

Trauma is the 5th leading cause of death in the United States and hemorrhage alone accounts for up to 56% of deaths within the first 24 hours of trauma. Exsanguination is the most common cause of death among those found dead upon the arrival of emergency medical services personnel. There is currently a lack of diagnostic technology at the point of injury leading to a standard of care that is less than optimal until a hospital or intensive care unit can be reached [1].

Appropriate triage of patients to higher and more advanced echelons of care will require smaller, more advanced technology that can be employed by first responders and caregivers. Fully integrated, rapidly deployable, minimally invasive biosensor systems for the continual measurement of metabolic markers of trauma, such as glucose and lactate, are at the forefront of diagnostic development to bridge this technological gap. The first generation of enzyme-based amperometric biosensors utilizes microchip technology that enables multi-analyte monitoring. One such biotransducer, the Electrochemical Cell-on-a-Chip Multidisc Electrode Array (MDEA 5037), has been developed and being used for the continual measurement of glucose and lactate in a vertebrate animal hemorrhage model [2]. The oxidoreductase enzymes glucose oxidase (GOx) and lactate oxidase (LOx) were immobilized onto MDEA 5037 transducer working electrodes via electropolymerization of a poly(pyrrole-*co*-pyrrole butyric acid) (p(Py-*co*-PyBA) copolymer. Electropolymerizations were performed at 850 mV *vs.* Ag/AgCl (3 M KCl) for a charge density of 100 mC/cm^2. Biotransducers prepared with a copolymer of a 7:1 ratio of pyrrole (Py) to pyrrole butyric acid (PyBA) were shown to have a sensitivity of 4.75 (±0.93) μA/mM/cm^2, a K_{Mapp} of 7.20 (±2.59) mM, an I_{max} of 76.75 (±25.50) μA/cm^2, a response time of 16.5 (±5) s, and a detection limit of 0.05 (±0.03) mM. The foregoing use of the p(Py-co-PyBA) co-polymer in fabricating biotransducers yielded a significant fivefold increase in sensitivity ($p < 0.001$) and a threefold increase in maximum current ($p < 0.01$) compared to biotransducers fabricated with polypyrrole (PPy) only. Biosensors were implanted into the trapezius muscle of Sprague-Dawley rats to measure interstitial analytes. The sensor was implanted for a total of 4 hours. After resection of the biotransducer and subsequent testing *in vitro* the device was shown to decrease in sensitivity and increase in its limit of detection by 86%. Response time of resected biotransducers was 438 s, more than 10 times greater than freshly prepared biotransducers. This is likely due to biofouling and/or protease related enzyme denaturation. For long duration implantable devices [3] an electroconductive hydrogel [4] will be employed to confer better antifouling properties.

References:

[1] Kotanen CN, Guiseppi-Elie A. Monitoring systems and quantitative measurement of biomolecules for the management of trauma. Biomed Microdevices. 2013;15:561-77.

[2] Guiseppi-Elie A. An implantable biochip to influence patient outcomes following trauma-induced hemorrhage. Anal Bioanal Chem. 2011;399:403-19.

[3] Karunwi O, Wilson AN, Kotanen C, Guiseppi-Elie A. Engineering the Abio-Bio Interface to Enable More than Moore in Functional Bioelectronics. Journal of The Electrochemical Society. 2013;160:B60-B5.

[4] Kotanen CN, Tlili C, Guiseppi-Elie A. Bioactive electroconductive hydrogels: The effects of electropolymerization charge density on the storage stability of an enzyme-based biosensor. Appl Biochem Biotechnol. 2012;166:878-88.

(413h) Molecular Recognition Using Nanotube-Adsorbed Polymer Complexes

Zhang, J., Massachusetts Institute of Technology
Barone, P. W., Massachusetts Institute of Technology
Kim, J. H., Hanyang University
Landry, M., Massachusetts Institute of Technology
Lin, S., Massachusetts Institute of Technology
Ulissi, Z., Massachusetts Institute of Technology
Lin, D., Massachusetts Institute of Technology
Mu, B., Georgia Institute of Technology
Heller, D. A., Massachusetts Institute of Technology
Boghossian, A. A., Massachusetts Institute of Technology
Hilmer, A. J., Massachusetts Institute of Technology
Rwei, A., Massachusetts Institute of Technology
Hinckley, A., Massachusetts Institute of Technology
Kruss, S., Massachusetts Institute of Technology
Shandell, M., Massachusetts Institute of Technology
Nair, N., University of Colorado
Blake, S., Massachusetts Institute of Technology
Sen, F., Massachusetts Institute of Technology
Sen, S., Massachusetts Institute of Technology
Croy, R., Massachusetts Institute of Technology
Li, D., Massachusetts Institute of Technology
Yum, K., Massachusetts Institute of Technology
Ahn, J. H., Massachusetts Institute of Technology
Jin, H., Massachusettes Institute of Technology
Essigmann, J., Massachusetts Institute of Technology
Blankschtein, D., Massachusetts Institute of Technology
Strano, M. S., Massachusetts Institute of Technology
Originally presented on: 11/6/2013 10:36:00 - 10:54:00

Molecular recognition is central to the design of therapeutics, chemical catalysis and sensor platforms, with the most common mechanisms involving biological structures such as antibodies[1] and aptamers[2,3]. The key to this molecular recognition is a folded and constrained heteropolymer pinned, via intra-molecular forces, into a unique three-dimensional orientation that creates a binding pocket or interface to recognize a specific molecule. An alternate approach to constraining a polymer in three-dimensional space involves adsorbing it onto a cylindrical nanotube surface[4-7]. To date, however, the molecular recognition potential of these structured, nanotube-associated complexes has been unexplored. In this work, we demonstrate three distinct examples in which synthetic polymers enable unique and highly selective molecular recognition once adsorbed onto a single-walled carbon nanotube (SWCNT) surface. The phenomenon is shown to be generic, with new recognition complexes demonstrated for riboflavin, l-thyroxine, and estradiol, predicted using a 2D thermodynamic model of surface interactions. The dissociation constants are continuously tunable by perturbing the chemical structure of the heteropolymer. The sensing mechanism is confirmed by simultaneous dual-channel single-molecule imaging of the SWCNT and the heteropolymer. The complexes can be used as new types of sensors based on modulation of SWCNT photoemission, as demonstrated using a complex for real time spatio-temporal detection of riboflavin in murine macrophages.

References

1 Scanlan, C., Offer, J., Zitzmann, N. & Dwek, R. Exploiting the defensive sugars of HIV-1 for drug and vaccine design. *Nature* **446**, 1038-1045 (2007).

2 Ellington, A. D. & Szostak, J. W. In vitro selection of RNA molecules that bind specific ligands. *Nature* **346**, 818-822 (1990).

3 Cho, E. J., Lee, J. W. & Ellington, A. D. Applications of Aptamers as Sensors. *Annu. Rev. Anal. Chem.* **2**, 241-264 (2009).

4 Nish, A., Hwang, J., Doig, J. & Nicholas, R. Highly selective dispersion of single-walled carbon nanotubes using aromatic polymers. *Nature Nanotechnology* **2**, 640-646 (2007).

5 Tsyboulski, D. *et al.* Self-Assembling Peptide Coatings Designed for Highly Luminescent Suspension of Single-Walled Carbon Nanotubes. *J. Am. Chem. Soc* **130**, 17134-17140 (2008).

6 Zheng, M. *et al.* DNA-assisted dispersion and separation of carbon nanotubes. *Nature Materials* **2**, 338-342 (2003).

7 Tu, X., Manohar, S., Jagota, A. & Zheng, M. DNA sequence motifs for structure-specific recognition and separation of carbon nanotubes. *Nature* **460**, 250-253 (2009).

(471a) Simulation and Experimental Investigation of a SAW Sensor With Delay Path Modifications

Richardson, M. B., University of South Florida
Koochakzadeh, S., University of South Florida
Suthar, K., Argonne National Laboratory
Sankaranarayanan, S., Argonne National Laboratory
Bhethanabotla, V. R., University of South Florida
Originally presented on: 11/6/2013 12:30:00 - 12:48:00

We developed a surface acoustic wave (SAW) device that includes an array of cavities having square cross-sectional areas (microcavities). It is common when dealing with SAW sensors to have high insertion losses. Several approaches have been utilized which includes the incorporation of reflective gratings, grooves and corrugated surfaces, and the application of a waveguide material to remedy this problem. Decreasing the amount of power consumed will limit the amount of heat generated that can adversely affect protein activity and increasing sensitivity will allow the detection of antigens at clinically relevant concentrations. Two different substrates commonly used in biosensors were studied, ST-Quartz and 36° YX -$LiTaO_3$. From simulations, we find that microcavities optimized for a particular substrate and wave mode scatter the incident wave causing constructive interference while acting as a waveguide limiting its penetration into the bulk. Using results from the simulations as a basis we fabricated devices on both substrates that support shear-horizontally polarized waves. A standard SAW delay-line and a SAW device containing microcavities were compared by measuring insertion loss and determining the sensitivity to changing glycerol concentrations. Our experimental results agree well with simulations and we conclude that a delay path modification in the form of microcavities is an effective way to improve SAW sensing performance.

(471b) Biosensor Combining the Principles of Metal Enhanced Fluorescence and Surface Acoustic Waves (MEF-SAW)

Morrill, S.,
Bhethanabotla, V. R., University of South Florida
Dutta, D., University of South Florida
Richardson, M. B., University of South Florida
Originally presented on: 11/6/2013 12:48:00 - 13:06:00

We propose a novel biosensor that integrates the two principles of metal enhanced fluorescence (MEF) and surface acoustic waves (SAW) on a single chip. MEF enhances the optical sensing signal intensity, whereas, Rayleigh SAWs have previously been shown to remove non-specifically bound (NSB) proteins and enhance fluid mixing by the acoustic streaming phenomenon. Silver island films were sputtered on the delay path of split-finger IDTs on a $LiNbO_3$ chip. AFM images revealed the formation of varying size silver islands. By optimizing the sputtering conditions, some silver islands were bigger than 15 nm, resulting in greater fluorescence enhancement.

Though fluoro-immunoassay is a widely used tool in medical diagnostics, it suffers from certain photophysical constraints like low quantum yields, autofluorescence, and low photostability. However, these constraints can be addressed by modifying the spectral properties of the fluorophore using MEF, a physical phenomenon in which the fluorescence signal of an emitting fluorophore is considerably enhanced by the presence of vicinal metal nanoparticles. MEF immunoassay suffers from certain limitations like low signal-to-noise ratio due to spurious signals due to interference from non-specific binding proteins and longer detection times. However, these limitations of the MEF immunoassay technique are easily overcome by combining it with a Rayleigh wave SAW device, in which Rayleigh waves help in efficient droplet mixing of the fluid, thereby considerably reducing the detection time. Additionally, the SAW-based Rayleigh waves help in selective removal of NSB proteins, which is useful for renewing the surfaces of the biosensors and in eliminating biofouling.

Experiments were conducted to demonstrate the concepts of MEF, the removal of NSB proteins, and the efficient micro-mixing on the MEF-SAW device. The results demonstrate enhanced sensing due to the combined effects of MEF and SAW. This novel MEF-SAW device provides a unique platform for both detection and removal of biological antigens on the same substrate, with shorter detection times than standard fluoro-immunoassays, enhanced signal-to-noise ratio, and increased sensitivity.

(471c) Micron-Scale Retroreflectors As Labels for Sensitive Point-of-Care Diagnostics

Garvey, G., University of Houston
Shakrarisaz, D.,
Raja, B., University of Houston
Kar, A.,
Pascente, C.,
Ruchhoeft, P., University of Houston
Kourentzi, K., University of Houston
Willson, R. C., University of Houston
Originally presented on: 11/6/2013 13:06:00 - 13:24:00

Timely diagnosis and treatment of infectious diseases calls for highly sensitive, point-of-care diagnostic assays. Recent technological advances have enabled the development of such point-of-care tests including microfluidic chip sensors, paper-based fluidic devices and lateral flow assays (LFAs). Many of these diagnostic methods, however, require elaborate sample preparation and many of the most common labels (typically a dye, enzyme or fluor) can suffer from low signal strength or photo-bleaching (of fluors), and can require elaborate instrumentation for their detection.

Here, we propose a combination of diagnostic platforms utilizing micron-scale, lithographically fabricated corner cube retroreflectors and high refractive index barium titanate microspheres as ultra-bright, highly detectable, optical labels. The retroreflector cubes, and barium titanate spheres are easily detected using inexpensive, low-numerical aperture optics, are not susceptible to photo-bleaching and do not require any signal amplification. Furthermore, the detectable nature of these particles minimize or eliminate altogether labor intensive sample preparation processes required for complex matrices.

We demonstrate the applicability of these novel assay formats by the two, and single-step, direct detection of pathogenic organisms and protein biomarkers in sub-milliliter volumes of various sample matrices.

(471d) Development of a DNA Methylation Sensor Based On Fluorescence Correlation Spectroscopy

Yuan, C., Purdue University
Kim, S. E., Purdue University
Originally presented on: 11/6/2013 13:24:00 - 13:42:00

Biomarkers based on DNA methylation patterns dictate the combined genetic (inherited) and epigenetic (environmental) manifestations of cancer risk and are highly promising to detect early-stage cancers, such as lung and prostate cancers. However, the current approaches for finding novel methylation targets and detect small changes in DNA methylation levels either do not provide sufficient resolution or are laborious to perform. We engineered a protein probe based on DNA CpG methyl-binding-domains to overcome these challenges. Coupled with the fluorescence correlation spectroscopy, the engineered protein probe can be used to detect small changes in DNA methylation level using low concentrations of chromosomal DNA, i.e., ~pM. The sensitivity and specificity of the engineered probe are characterized. Extension of this platform for *in vivo* detection of DNA methylation patterns will also be discussed.

(471e) Equilibrium and Dynamics of Human IgG Adsorption to a Peptide Affinity Ligand On Pure and Mixed Monolayers

Islam, N.,
Gurgel, P., North Carolina State University
Rojas, O. J., North Carolina State University
Carbonell, R. G., Kenan Institute for Engineering, Technology & Science, NC State University
Originally presented on: 11/6/2013 13:42:00 - 14:00:00

In this study, the behavior of a hexameric affinity peptide was characterized in depth and its potential as a means of immunoglobulin G (IgG) detection was explored. Surface Plasmon Resonance (SPR) was used to study the binding of human IgG to the hexameric peptide affinity ligand HWRGWV, which is covalently grafted to pure or mixed alkanethiol self-assembled monolayers (SAM) on gold surfaces. The reactive alkanethiol ($HS\text{-}(CH_2)_{11}\text{-}(CH_2CH_2O)_6\text{-}NH_2$, shortened notation, EG_6NH_2) is $-NH_2$ terminated, contains six ethylene glycol units, and was used to create pure monolayers on gold sensor surfaces. The NH_2 termini are reactive towards the C-terminus of the peptide ligands. To prepare mixed monolayers with the above thiol, a 'dilutor' thiol ($HS\text{-}(CH_2)_{11}\text{-}(CH_2CH_2O)_3\text{-}OH$, shortened notation, EG_3OH) with three ethylene glycol units and a hydroxyl terminus group, was utilized. The thiols were mixed at various percentages of EG_6NH_2 (with 0% denoting pure EG_3OH, 100% denoting pure EG_6NH_2) to form monolayers of varying NH_2densities on gold sensors and slides.

The pure or mixed monolayer surface density and peptide grafting density were quantified using Time of Flight Secondary Ion Mass Spectrometry (ToF SIMS) and ellipsometric measurements. The absence of phase islands of mixed monolayers was verified with contact angle measurements and the successful grafting of the peptide on these monolayers was also verified through ToF-SIMS. Peptides were covalently grafted via their C-termini to produce active SPR sensors and various protein binding experiments were carried out with SPR. The dynamics and extent of peptide-IgG binding were compared with results from conventional systems based on Protein A attached on gold via disulfide monolayers. Equilibrium and kinetic binding parameters for IgG were obtained and compared, for peptide ligands grafted on to pure and mixed monolayer systems, in order to make a comprehensive study of the roles of N-terminal capping, of peptide density and of the base matrix components on the IgG-peptide binding. Non specific binding was investigated by BSA binding experiments. The behavior of complex mixtures (Chinese Hamster Ovary supernatant, or CHO, containing IgG and other impurities) on the peptide sensors was studied, and was compared with pure IgG binding behavior. Different regeneration buffers were tested to obtain the optimum buffer, providing maximum regeneration, without affecting the peptide performance. The grafting density and the fractional conversion of NH_2 moeities of the monolayer to peptides, do not seem to play a significant role in the capacity or affinity of IgG binding, but the ratio of active thiol EG_6NH_2 to dilutor thiol EG_3OH

has an effect on the selectivity of binding. The optimum selectivity of IgG to BSA binding was obtained with peptides attached to mixed monolayers formed from 10% EG_6NH_2-90% EG_3OH thiol solution in ethanol. IgG binding to acetylated HWRGWV (Ac-HWRGWV) supported on 100% EG_6NH_2 SAM showed an equilibrium dissociation constant, K_d of 5.8x 10^{-7} M and that for HWRGWV on the 100% EG_6NH_2 SAM (after deprotection of Fmoc-HWRGWVA) yielded K_d of 1.2×10^{-6} M. On the peptide with mixed monolayer system the K_d was 9.33×10^{-7} M. Maximum binding capacities, Q_m were 3.8, 4.2 and 3.2 mg m^{-2} for 100% EG_6NH_2-Ac-HWRGWV, 100% EG_6NH_2-HWRGWV and 10% EG_6NH_2-Ac-HWRGWV respectively. Real-time SPR data were analyzed with mass transport effects taken into consideration and were used to determine the apparent rate constants for association and all three peptide based systems yielded similar values for k_a (2.2, 2.1 and 1.9 m^3 mol^{-1} s^{-1}, respectively). In binding experiments for IgG from complex mixtures, it was observed that that for diluter mixtures, the mixed monolayer system showed better performance, while for complex mixtures with higher IgG content, the binding results with pure monolayer + peptides conformed better with that of pure IgG mixture solutions. The regeneration-binding cycles for 100% EG_6NH_2-Ac-HWRGWV as well as 10% EG_6NH_2- Ac-HWRGWV systems showed optimum performance with 10% acetonitrile mixed with 0.1 N NaOH. At least 8 cycles of regeneration and binding were demonstrated with little or no change in performance of the peptides, both on pure and mixed monolayers, with this regeneration buffer. The experiments carried out in the study helped understand the mechanism of binding of IgG to the small molecule ligands. These and previous experimental results by Yang et al working with HWRGWV peptide[1, 2]indicate that neither hydrophobic nor electrostatic interactions are a dominant force in binding and that hydrogen bonding between residues play a major role in the specific affinity binding. The results of regeneration experiments carried out in this study indicate that merely the change of pH is not enough to cause 100% breaking of bonds between IgG and the peptide, as seems to be the case for Protein A regeneration. An introduction of a chaotropic agent (eg urea) or a polar solvent (eg. acetonitrile) is required to weaken forces contributed by hydrogen bonding. Results of this study can help formulate an effective biosensor for hIgG, and also helps further understand the mechanism of binding between hIgG and small molecule affinity ligand HWRGWV.

1. Yang, H. O.; Gurgel, P. V.; Williams, D. K.; Bobay, B. G.; Cavanagh, J.; Muddiman, D. C.; Carbonell, R. G., Binding site on human immunoglobulin G for the affinity ligand HWRGWV. *Journal of Molecular Recognition* **2010,**23, (3), 271-282.

2. Yang, H.; Gurgel, P. V.; Carbonell, R. G., Purification of human immunoglobulin G via Fc-specific small peptide ligand affinity chromatography. *Journal of Chromatography A* **2009,** 1216, (6), 910-918.

(471f) Molecular Combing of Oligonucleotides Induces Secondary Structure Specific Azimuthal Liquid Crystal Orientation

Noonan, P. S., University of Colorado Boulder
Monserud, J. H., University of Colorado Boulder
Schwartz, D. K., University of Colorado Boulder
Originally presented on: 11/6/2013 14:00:00 - 14:18:00

Since its advent in the 1990s, molecular combing of DNA has been exploited for DNA sequencing, gene expression mapping, and pathogen detection. To date, only long (>kbp) DNA strands have been used in these strategies, presumably due to inadequate methods for characterizing the elongation of short oligonucleotides. Previous work in our group using λ-DNA revealed DNA structure specific azimuthal liquid crystal (LC) orientations at near single-molecule sensitivity. In the current work we demonstrate that a similar phenomenon is observed when using short (<20 nucleotides) oligonucleotides. Amine-terminated ssDNA was immobilized onto a (3-glycidoxypropyl) trimethoxysilane modified substrate and stretched by introducing a receding meniscus in a controlled direction. We found that immobilized ssDNA induced azimuthal LC orientation with the LC molecules aligned in the stretching direction. When we introduced a complementary ssDNA molecule to the aqueous phase prior to stretching, hybridization occurred and the resulting azimuthal orientation was rotated 30° from the stretching direction. In order to demonstrate the potential of this approach for biosensing applications, a microfluidic device was also used to introduce the receding meniscus. Correlation was observed between the shape of the receding air-water interface and the azimuthal LC orientation indicating the expected DNA stretching perpendicular to the interface. This is the first demonstration where molecular combing was used to stretch short oligonucleotides on a solid substrate in a uniform direction, thus illustrating the potential for a highly sensitive DNA hybridization detection strategy that can readily be developed into a multiplexed assay.

(471g) Single-Particle Assay for Detecting Influenza Virus and Characterizing Its Binding and Fusion Rates

Lee, D., Cornell
Hsu, H. L.,
Daniel, S., Cornell University
Originally presented on: 11/6/2013 14:18:00 - 14:36:00

The rapid detection and characterization of enveloped viruses are important for controlling viral pandemics. Here we discuss a single-particle technique for detecting influenza through receptor binding and membrane fusion behaviors. Influenza utilizes surface proteins called hemagglutinin to bind to sialic acid groups on the host cell surface, and to initiate the fusion of the viral and host membrane. Membrane fusion is triggered at a low pH and is required for infection. For the assay, we use a microfluidic device made on a glass substrate that has been coated with a supported lipid bilayer (SLB). The SLB contains sialic acid receptor molecules and can be modified to mimic various host cell membranes. Total internal reflection microscopy is used to monitor the SLB and any fluorescent object that interacts with it. To detect the viral envelope, trace amounts of lipophilic fluorophores were added to the virus-contaminated solution before loading it into the device. We were able to detect individual virions attaching to the SLB, and more importantly, quantify the rate at which binding and unbinding occurs for each virion. To detect membrane fusion activity, an acidic solution was flown over the bound viruses. The spread of the lipophilic fluorophore from the viral membrane to the SLB confirmed fusion activity and the rate at which this occurs was quantified. Using the binding, unbinding, and fusion rate information, one could potentially produce a list of host species that are susceptible to infection and thus implement appropriate quarantine procedures.

(471h) Three Stage Sample Preparation for Biotoxin Detection in Blood

Javanmard, M., Stanford Genome Technology Center
Emaminejad, S., Stanford University
Gupta, C., Stanford University
Davis, R., Stanford Genome Technology Center
Howe, R. T., Stanford University
Originally presented on: 11/6/2013 14:36:00 - 14:54:00

The use of spectroscopic based techniques such as raman spectroscopy or quantum tunneling spectroscopy (Gupta et. al) for analyzing low abundance proteins in blood is difficult due to the high level of background arising from the presence of blood cells and also highly abundant proteins (~1-1000μM) like albumin, IgG, IgA, transferrin, haptoglobin and α-1-antitrypsin making up 90% of the protein mass. The dynamic range of proteins in blood plasma alone spans nine orders of magnitude resulting in difficulties for quantification. Also, depletion of cells from blood requires centrifugation which is bulky and unsuitable for point of care diagnostics. Finally, the orientation at which the proteins are immobilized on the surface of the sensor also will affect the consistency of the sensor reading and also the signal to noise ratio. We present a three stage on-chip platform for depletion of cells and highly abundant serum proteins in blood, and finally controlling the orientation at which the target proteins become immobilized to the sensor surface. Our platform consists of three components, the first of which is a microfluidic mixer which mixes beads containing antibodies against the highly abundant proteins with the whole blood. For microfluidic mixing we use an adapted herringbone structure (Whitesides et. al) to induce chaotic mixing of beads with proteins in the channel. Using the adapted herringbone serpentine channel we achieve depletion of Immunoglobin G all the way down to 35 nanoMolar in concentration.

This complex mixture (consiting of beads, cells, and serum proteins) is then injected into the second component of our microfluidic platform which consists of a filter trench to capture all of the cells and the beads. In order to significantly enhance the trapping of the cells and beads into the filter trench we have integrated a pair of interdigitated electrodes at the top of the channel above the trench in order to use negative dielectrophoresis in order to push all of the micron sized particles (cells and beads which have captured the highly abundant proteins) down into the trench allowing the serum proteins of lower abundance to flow through. In general dielectrophoresis is incapable of producing forces beyond the low piconewton range. However using atomic layer deposition to deposit a 10 nm pin-hole free layer of SiO2 on top of the electrodes protecting them from corrosion, we were able to demonstrate DEP forces in the nanonewton range. Using the enhanced negative DEP electrodes in combination with the filter trench, we were able to demonstrate 100% depletion of all micron sized particles in the mixture.

The third component involves a novel method for tunably controlling the immoblization of proteins on a solid-state surface using electric field. We exploit the dipole property of IgG and its ability to be oriented with field. As a proof of concept, we use fluorescence detection to indirectly verify the modulation of the orientation of proteins bound to the surface. We studied the interaction of fluorescently tagged anti-IgG with surface immobilized IgG controlled by electric field. Our study demonstrates that the use of electric field can result in up to 40X enhancement in signal to noise ratio compared to normal physical adsorption.

(530a) Molecularly - Imprinted Based Sensors for Real-Time and Accurate Monitoring of Volatile Organic Compounds

Deng, Y., Arizona State University
Chen, C., Arizona State University
Tsow, F., Arizona state university
Xian, X., Arizona State University
Forzani, E., Arizona state university
Originally presented on: 11/6/2013 15:15:00 - 15:31:00

Molecularly - imprinted based sensors for real-time and accurate monitoring of volatile organic compounds

Yue Deng,[1,2] Cheng chen,[1,2] Francis Tsow,[1] Xiaojun Xian,[1] Erica Forzani,[1,2*]

[1]Center for Bioelectronics and Biosensors, Biodesign Institute; [2]Ira A. Fulton Schools of Engineering, Arizona State University, AZ,

***eforzani@asu.edu**

Abstract: Volatile organic compounds(VOCs) are organic chemicals that have a high vapor pressure at ordinary, room-temperature conditions. Some VOCs are dangerous to human health or cause harm to the environment. To improve our understanding of indoor and outdoor personal exposures to common environmental toxicants daily released into the environment, new technologies that can monitor and quantify the toxicants anytime anywhere are needed. This presentation presents a wearable sensor to provide such capabilities. The sensing material is molecularly imprinted polymer, which is highly selective. The sensor combines the polymer and micro fabricated quartz tuning fork detector, it can communicate with a common cell phone through Bluetooth, and allows for accurate measurement of volatile organic compounds at personal level in real-time. Stable performance was validated using Gas Chromatography - Mass Spectrometry (GC-MS) and Selected Ion Flow Tube -Mass Spectrometry (SIFT-MS) reference methods in a variety of environments and activities. Field tests were carried out to examine personal exposure in various scenarios including: indoor and outdoor environments, traffic exposure and personal exposure in different cities, countries, and near the 2010 Deep-water Horizon's oil spill. These field tests not only validate the performance, but also demonstrate the unprecedented high temporal and spatial toxicant information provided by the new technology.

(530b) A Microplasma Glow Discharge For Fire Detection

Vander Wal, R. L., Pennsylvania State University
Gaddam, C. K., Pennsylvania State University

Originally presented on: 11/6/2013 15:31:00 - 15:47:00

Abstract

In this study the main objective was to develop and demonstrate a glow discharge microplasma coupled to a miniature spectrometer for detection of fire signatures from pyrolyzing and burning spacecraft materials. Our experimental results demonstrate that combustion-produced carbonaceous aerosols can serve to identify the burning materials. Demonstrating versatility for chemistry analysis, the plasma detector could differentiate carbonaceous aerosols with different C/H ratios and distinguish inorganic samples such as salts and metal oxides from carbonaceous aerosols. In addition, in-situ analysis of aerosol samples validated the microplasma's analytical utility by linearity of its optical emission intensity with aerosol elemental composition.

Introduction

Plasmas are used as direct analytical tools for optical emission spectroscopy. ICP (inductively coupled plasma) [1] and DCP (direct-current plasma) [2] instruments create high temperature plasmas that perform the dual analytical function of atomizing and electronically exciting the constituent elements whose spectra uniquely identify the material composition. Liquids are readily analyzed and even solids, if sample digestion is first performed using strong acids. Glow discharge plasmas are also used in commercial laboratory instruments for atomization of solid samples. Typically a kW power supply powers the glow discharge under a low-pressure inert gas to atomize near surface layers of a material for subsequent mass spectrometry analysis [3]. Nevertheless, mass spectrometry is difficult to calibrate, particularly over a wide range of atomic masses.

In recent years the hollow cathode glow discharge has been miniaturized [4] thereby testing physical scaling relationships. Specifically these relate the operating pressure to the relevant (hollow cathode) physical dimension, i.e. the "hollow" diameter [5]. By following White's relation [6], atmospheric pressure operation could be achieved with hollow cathode spatial dimensions on the order of 100 microns diameter. Significantly the need for low-pressure operation and vacuum hardware [7] is negated. A second advantage is the drastic reduction of operating power. Both factors greatly simplify the supporting hardware. Single and arrayed elements have been successfully demonstrated [8]. To-date, research of such microdischarges has focused upon their applications to photonics and their use as VUV light sources [9].

Uniquely, in this work we synergize several of these concepts towards a micro-plasma fire detector. Though other microplasmas have been developed and demonstrated for detecting gases [10-12], their geometry is not well suited for analysis of solids or particulate. Relative to flowing configurations, the micro hollow glow discharge (MHGD) offers unique capability for analysis of combustion-produced particulate. Despite recent reviews, microplasma application to detection and analysis of aerosols, suspended or deposited, was not identified, though detailed studies of traditional glow discharges applied to their analysis as deposit, and for that matter, more generally to powder and solids.

Using recent advancements in micro-hollow cathode technology for operation at atmospheric pressure, we utilize the traditional sputtering capability of the glow discharge to both atomize and electronically excite samples for stand-alone analytical analysis, akin to dedicated laboratory scale instruments but now on a microscale with Watt-level power usage. Moreover we analyze carbonaceous materials, notorious for their traditional analytical challenges not least of which is sample atomization/digestion and light element detection, in particular hydrogen content, quantitatively. Combustion produced soot can vary widely in composition, in particular C/H ratio, reflecting both its molecular origins and temperature of evolution [13,14]. Such samples are of interest for a variety of reasons ranging from atmospheric aerosols and their impact upon global warming [15], health impacts associated with particulate emissions from combustion sources [16] and finally as potential fires signatures [17].

One of the primary goals of the fire detection element of NASA's Fire Safety Program is to provide recommendations for the development of more effective and efficient fire sensors to be installed in space vehicles and future habitats. Detectors that can detect minute amounts of soot and additionally analyze its composition could not only provide early warning but also act as a "smart detector" by identifying the material producing the soot, whether in smoldering or flaming combustion. Clearly low detection limits are desirable. Moreover, false-positives, "detection" of inorganic mineral matter leading to a fire alarm must be minimized given their disruptive nature. To satisfy these dual constraints a multifunctional glow discharge analyzer was developed to not only detect and differentiate soot particles generated from the combustion of spacecraft materials but also to distinguish them from inorganic samples such as salts and metal oxides.

Material and Methods

Materials approved for use aboard spacecraft were subjected to pyrolysis and combustion in a combustion system described elsewhere [18]. Concisely, the combustion system consists of a tubular electric furnace mounted vertically and fitted with a quartz tube. The furnace is equipped with a temperature controller and the quartz tube is connected to different gas lines. The sample mass varied, depending on the particulate yield from the material. The combustion tests for all samples were performed at 1000 ^{o}C with an air flow of 150 sccm, upwards. Soot and condensables were collected on a copper disc of 0.127 mm thickness and then analyzed in the micro-hollow glow discharge. Materials evaluated included polypropylene pellets, polyurethane foam, silicone rubber, Kapton sheet, Teflon and Halar wire insulations that conform to current NASA flammability standards as outlined in by the 6001D Materials Outgasing and Flammability test [19].

Conclusion

Based on our studies we have demonstrated that the glow-discharge, plasma-based detection system is capable of detecting and identifying soots produced from combustion of six materials approved for use in space vehicles and habitats. Based upon the distinctiveness of the spectra, the (material) source of the fire event can be identified by analyzing the MHGD plasma emission spectrum of the corresponding soot. In addition, the micro-plasma analysis method is able to determine the C/H elemental ratio of carbonaceous aerosols and distinguish these combustion-produced emissions from ambient inorganic (nuisance) aerosols. Our detector is highly specific to smoldering or combustion events (by selective detection of emission products) and can reduce if not eliminate false alarms due to ambient dust and detritus. Finally, we have successfully demonstrated a linear correlation between the spectral intensity versus sample mass, further extending the analytical capabilities of the MHGD plasma instrument as a "smart" fire sensor.

References

[1] P. W. J. M Boumans, Inductively Coupled Plasma Emission Spectroscopy, John Wiley and Sons, New York, 1987, Part I.

[2] W. Lochte-Holtgreven, "Plasma Diagnostics", Amsterdam: North-Holland, 1968.

[3] W. Schelles, K. J. R. Maes, S. DeGent, R. E. V. Grieken, Anal. Chem. 68, (1996) 1136-1142.

[4] J. G. Eden, S.-J. Park, N. P. Ostrom, K.-F. Chen, K.-F., J. Phys. D: Appl. Phys. 38 (2005) 1644-1648.

[5] K. H. Schoenbach, R. Verhappen, T. Tessnow, F. E. Peterkin, W. W. Byszewski, Appl. Phys. Lett. 6 (1996) 13-15.

[6] A. D. White, J. Appl. Phys. 30, (1959)711-719.

[7] A. El-Habachi, K. H. Schoenbach, Appl. Phys. Lett. 72 (1998) 22-24.

[8] O. Sakai, S. Hashimoto, A. Hatano, Appl. Phys. Lett. 82 (2003) 2781-2783.

[9] M. Moselhy, R. H. Stark, K. H. Schoenbach, U. Kogelschatz, Appl. Phys. Lett. 78 (2001) 880-882.

[10] V. Karanassios, Spectrochim. Acta Part B 59 (2004) 909-928.

[11] M. Miclea, M. Okruss, K. Kunze, N. Ahlman, J. Franzke, Anal. Bioanal. Chem. 388 (2007) 1565-1572.

[12] J. A. C. Broekaert, Anal. Bioanal. Chem. 374 (2002) 182-187.

[13] M. Alfè, B. Apicella, J.-N. Rouzaud, A. Tregrossi, A. Ciajolo, Combustion and Flame, 157 (2010) 1959-1965.

[14] R. L. Vander Wal, Combustion and Flame 11 (1998) 607-616.

[15] R. J. Charlson, S. E. Schwartz, J. M. Hales, R. D. Cess, J. A. Coakley, Jr., J. E. Hansen, D. J. Hofmann, Science 255 (1992) 423-430.

[16] J. S. Lighty, J. M. Veranth, A. F. Sarofim, Journal of Air and Waste Management 50 (2000) 1565-1618.

[17] C. Forster, U. Wandinger, G. Wotawa, P. James, I. Mattis, D. Althausen, P. Simmonds, S. O'Doherty, S. G. Hennigs, C. Kleefeld, J. Schneider, T. Trickl, S. Kreipl, H. Jager, A. Sohl, J. Geophys. Res., 106 (2001) 887-22906.

[18] R. L. Vander Wal, V. Pushkarev, J. H. Fujiyama-Novak, "Fire Signature of Spacecraft Materials: Gases and Particulates", Combustion and Flame, in press.

[19] D. R. Mulville, Flammability, Odor, Offgassing and Compatibility Requirements and Test Procedures for Materials in Environments that Support Combustion (1998) NASA-STD-6001.

(530c) Development of An Impedance-Based Sensor for Detection of Catalyst Coking in Fuel Reforming Systems

Wheeler, J. L.,
Sullivan, N., Colorado School of Mines
Porter, J. M., Colorado School of Mines
Originally presented on: 11/6/2013 15:47:00 - 16:03:00

A novel sensor for detecting the early stages of catalyst coking in fuel reforming systems has been developed at the Colorado School of Mines. The sensor was manufactured by inkjet printing a colloidal suspension of ceramic powders to create thin (~20 μm) catalytic and conductive elements of the sensor. The catalytic elements of the sensor are composed of a Ni-YSZ cermet. The Ni-YSZ cermet was prepared with a concentration below the percolation limit (~20 vol%) of nickel, ensuring a low electrical conductivity. As coke forms on the catalyst material, the nickel nodules in the Ni-YSZ are connected by electrically conductive carbon and the conductivity of the catalyst material increases.

Sensors were tested in 1% ethylene and methane dry reforming environments to induce coking. The sensor showed a strong response to coking by producing a signal on the order of hundreds of millivolts. The mass of the coke load was determined to be below the detection limit of available thermogravimetric analyzers (TGA) (<10 μg). The coke load was further examined with a field emmision scanning electron microscope (FESEM) and was found to be primarily carbon nanofibers. Carbon nanofibers ~10-100 nm in diameter connected nickel nodules in the sensors' catalyst material resulting in a change in resistance in the catalyst material.

Tests were also carried out to determine if the sensor was capable of automatically controlling a fuel stream. Custom electronics were developed to interface the sensor to an Arduino UNO microcontroller. The Arduino microcontroller was able to read signals from the sensor and control the experimental fuel stream. Sensors were exposed to a methane dry reforming fuel stream (23% methane, 77% carbon dioxide at 640 °C) to induce coking. The sensor automatically shut down the coking fuel stream when a coking signal was detected, ~130 s after the sensor was exposed to coking conditions.

(530d) 'smart' Sand for Buried Explosive Detection By Naked Eye Under Handheld UV Light

Sun, X., University of Connecticut
Mopidevi, S., Department of Biomedical Engineering
Liu, Y., University of Connecticut
Silhavy, C., University of Connecticut
Lei, Y., University of Connecticut
Nieh, M. P., University of Connecticut
Originally presented on: 11/6/2013 16:03:00 - 16:19:00

In this study, explosive-sensitive fluorescent sand ('smart' sand) was prepared through simple coating technology. The properties of the as-prepared 'smart' sand were characterized by SEM, EDS, XRD, FT-IR, UV-vis and fluorescence emission spectra. The characterization results indicated that fluorescent explosive-sensing composite film was successfully coated on the surface of sand, and thus the 'smart' sand displayed strong fluorescence. The as-prepared 'smart' sand was further applied for the detection of buried explosives. Different amounts of 2,4-DNT were buried in soils with different depth. Then the 'smart' sand was 'sprayed' on the soil surface. Under handheld UV light, the fluorescence of 'smart' sand is strongly quenched by leaked explosive vapors, which can be visualized by naked eye. In addition, emission spectra of quenching spot and non-quenching spot after detection were collected to corroborate the quenching mechanism. This result indicates that 'smart' sand has the potential to be applied for large-scale landmine detection and mapping.

Sensors

(530e) In-Situ Synthesis of 3-D Nanostructures Within a Microfluidic Device and Their Applications As SERS Substrates

Parisi, J. Jr., University of Connecticut

Su, L., University of Connecticut

Lei, Y., University of Connecticut

Originally presented on: 11/6/2013 16:19:00 - 16:35:00

In-situ synthesis of 3-D nanostructures within a microfluidic device and their applications as SERS Substrates

Joseph Parisi, Liang Su, Yixin Liu, Yu Lei

Department of Chemical, Materials, & Biomolecular Engineering, University of Connecticut, 191 Auditorium Road, Unit 3222, Storrs, Connecticut 06269

Microfluidic devices have been used in many different applications and have continued to display their ability to shrink large scale experiments into small scale devices. In this report, a microfluidic device with integrated novel silver nanoparticle (Ag NPs) decorated nanowall structures was fabricated via *in-situ* electrodeposition of Cu-core/C-sheath nanowalls, followed by a facile in-channel silver galvanic replacement reaction method at room temperature. These nanostructures are fabricated in a fraction of the time compared to similar CVD and hydrothermal structures and require much less reagents and complicated fabrication steps. The integrated microfluidic devices with Ag NPs decorated nanowalls, serving as a highly active Raman substrate, was then applied for in-channel surface-enhanced Raman scattering (SERS) sensing. Single molecule detection is obtained using model compounds, and the detection of water contaminants is also presented. Because preparation of the active SERS substrate and subsequent finger-print SERS detection is completed within a microfluidic device, the developed method opens a new venue to integrate active SERS substrate within microfluidic channels and provides an excellent microfluidic SERS sensor platform for ultrasensitive and selective chemical and biological sensing.

(530f) Biomodular High Temperature Planar Oxygen Gas Sensor

Sun, X., University of Connecticut
Liu, Y., University of Connecticut
Lei, Y., University of Connecticut
Gao, P., University of Connecticut
Originally presented on: 11/6/2013 16:35:00 - 16:51:00

A planar O_2 bimodular sensor was fabricated by using yttria-stabilized zirconia (YSZ) substrate and NiO thin film. The NiO films were prepared by r.f. magnetron sputtering of NiO on YSZ substrate. The surface morphology of NiO film after sintering was observed by scanning electron microscopy (SEM) and atomic force microscopy (AFM). X-ray diffraction patterns suggested that NiO was coated on YSZ surface and it was stable even after high temperature oxygen sensing. The oxygen detection experiments were performed at 500-800°C under both potentiometric mode and resistance mode. For the potentiometric mode, the linear relationship between the EMF output of the sensor and the logarithm of the O_2 concentration was observed. For the resistance mode, the sensitivity increased with temperatures as well as concentrations, and the result was in good agreement with the relationship between the semiconductor electrical conductivity and oxygen concentrations. Both detection modes showed sensitive, reproducible and reversible response to oxygen. The combination of two detection modes into one sensor could open a new venue in high temperature gas sensing field.

(530g) Theoretical and Experimental Study of Modulation of the Fermi Level of CuPc/HOPG Via NO Adsorption

Choudhury, P., University of Pittsburgh
Park, J. H., UCSD
Kummel, A., UCSD
Originally presented on: 11/6/2013 16:51:00 - 17:07:00

Metal phthalocyanines (MPc) have been widely used for chemical vapor sensing due to their novel gas adsorption properties. In this work, we present both theoretical and experimental study of copper phthalocyanine (CuPc) interactions with a nonreactive substrate, HOPG, showing ground-state electronic modulation of the substrate near the Fermi level. In order to study the NO interactions onto CuPc molecules, we performed both van der Waals corrected spin-polarized periodic density functional theory (DFT) calculations and site specific scanning tunneling spectroscopy (STS) experiments. Both calculations and experimental results suggest that NO interacts weakly (physisorption) with CuPc/HOPG surface. The low coverage of NO adsorbates on HOPG in STM images is consistent with the small perturbation of neighboring density of states of CuPc. DFT calculations show that as NO adsorbs on metal center of CuPc, negative charge carriers are injected from CuPc molecules to NO oxidant. During this charge transfer process, excess electrons can be transfer to CuPc from the underneath layer of the substrate, HOPG, and HOPG substrates may be doped to p-type. The binding of NO to CuPc/HOPG also leads to the excess opening of the band gap by 0.18 eV as compare to the experimental value of 0.25 eV. The results of this work have shown that CuPc/HOPG could be a potential sensor device for oxidative analytes.

(530h) Impedancemetric High Temperature Gas Sensor Based On Electrospun Metal Oxide Nanofibers for Selective Reducing Gas Detection

Liu, Y., University of Connecticut
Lei, Y., University of Connecticut
Originally presented on: 11/6/2013 17:07:00 - 17:23:00

A high temperature gas sensor was developed based on electrospun metal oxide nanofibers. The morphology of the nanofibers was investigated by scanning electron microscopy and transmission electron microscopy. The impedance spectroscopy of the as-prepared nanofibers based sensor was investigated in N_2 and different gas mixtures (O_2, CO, C_3H_8, CO_2, NO and SO_2 balanced by N_2 in varied concentrations) at 800 ^oC. The sensing mechanism was investigated by equivalent circuit analysis. For the first time, the sensor was operated at high frequency, in which case strong reducing gas (CO and C_3H_8) can be selectively detected with good sensitivity, and the responses from other gases can be fully eliminated. These good results suggest that the as-prepared nanofibers is a promising material for the application of high temperature sensors for reducing gas detection and impedancemetric technique is a promising approach to improve the selectivity of gas sensors by choosing an appropriate applied frequency.

(530i) In-Situ Surface Plasmon Resonance Study On Methanol Oxidation On Platinum Nanoparticles

Choi, B.,
Jiang, P., University of Florida
Originally presented on: 11/6/2013 17:23:00 - 17:39:00

Surface plasmon resonance (SPR) is recognized as a label-free and direct technique for sensitive chemical and biological sensing. Here we present an in-situ SPR study on electrochemical heterogeneous catalysis in order to evaluate the possible use of a novel plasmonic substrate in catalytic mechanism studies with detection of reaction intermediates. The SPR substrate used in the study is prepared by a simple soft-lithography process using recordable digital versatile disc (DVD-R) as templates. Anatase titania or other metal oxide gratings can be easily replicated by using sol-gel processes. A 50 nm thick gold is finally sputtered on the templated grating. The values of the SPR sensitivity and the figure of merit (FOM) is determined to be 737 nm per refractive index unit (nm/RIU) and 73.1, respectively. For further electrochemical methanol oxidation study, platinum nanoparticles are electrochemically deposited on gold surface. Constant step voltage is applied from 0.00 to 1.00 V with 0.05 V of increment voltage in nitrogen purged methanol containing sulfuric acid electrolyte solution (0.05 M H_2SO_4 + 2.00 M CH_3OH) with chronoamperometry (CA) while optical probe is equipped to investigate the resonance peak position along the voltage change. Significant elevated peak shift (about 8 nm) is observed especially in low voltage region (from 0.00 to 0.55 V).

(627a) Controlling the Sensitivity of Integrated Silica Based Biosensors Through Surface Chemistry

Meinke, E., University of Southern California
Sun, V.,
Mehrabani, S., University of Southern California
Armani, A. M., University of Southern California
Originally presented on: 11/7/2013 8:30:00 - 8:45:00

Significant advances have been made in sensor device engineering and optimization, resulting in numerous high sensitivity transducers. While the limit of detection which is the most frequently reported metric is important, there are numerous other performance characteristics of equal importance, including the device working range. However, complementary research into methods to effectively control or tune the working range has not kept pace.

Recently, a method for covalently attaching targeting molecules to optical resonant microcavities was developed[1]. Using this approach, it is possible to both maintain the sensing performance of the microcavity and form dense, uniform layers of binding sites. However, a dense layer is not always desired, as the binding sites can interfere with each other. Additionally, the number of binding sites on the surface plays a role in determining the overall sensing threshold and detection response time. Therefore, by developing methods to tune the density of the binding sites, it will be possible to change the device performance.

In the present work, the number of active sites on the surface of integrated silica microcavity sensors is controlled by introducing a secondary "blocker" molecule, which is attached in parallel with the binding site. By changing the relative concentrations of the binding site and the null site, the overall density of the active sites on the surface can be controlled without introducing vacancies in the functionalization. In this proof of concept demonstration, two different attachment strategies are explored: one is based on the conventional NHS-ester method[1] and the second is based on the click chemistry (azide-alkyne) approach. Previous work combining PEG layers and microcavity sensors has demonstrated that PEG is an effective method for blocking non-specific binding[2]. By using two different attachment methods, it is possible to attach a wide range of probe molecules (eg antibodies, biotin) using commercially available material, enabling the rapid transition of the developed protocols.

Experiments are performed on both control substrates and integrated optical sensors. The silica sensors are lithographically fabricated in large arrays on silicon substrates. To verify and optimize the changes in binding site density, both the probe molecules and the PEG chain are fluorescently labeled. The resulting surfaces are characterized using fluorescence microscopy, and the differential fluorescence intensity is measured. This change indicates the proportion of PEG:binding site. Ongoing work is exploring the sensor performance (detection limit, response and equilibrium time, working range) as a function of the binding site density using both strategies.

1. H. K. Hunt, C. Soteropulos, and A. M. Armani, "Bioconjugation Strategies for Microtoroidal Optical Resonators," Sensors. **10**, 9317-9336 (2010).

2. C. Soteropulos, K. M. Zurick, M. T. Bernards, H. K. Hunt, "Tailoring the protein adsorption properties of whispering gallery mode optical biosensors," Langmuir, **28**, 15743-15750 (2012).

(627b) Point-of-Care Assay for Tuberculosis: Rapid Detection Using Droplet-Based Microfluidics

Rosenfeld, L., Stanford University
Cheng, Y., Stanford University
Rao, J., Stanford University
Tang, S. K. Y., Stanford University
Originally presented on: 11/7/2013 8:45:00 - 9:00:00

Tuberculosis is one of the most deadly diseases that kills over one million people each year and infects one-third of the world's population. The disease is spread by infection with Mycobacterium tuberculosis (Mtb). Owing to its airborne transmission, early diagnosis is critical to the prevention and control of TB. Standard diagnostic methods, acid-fast smear from sputum, often do not become positive until after transmission occurs, which allows the spread of the disease. Culture-based techniques are more sensitive, but take weeks to obtain results because of the extremely slow growth rate of Mtb.

In this study a new method to detect indicator enzyme based on the isolation of tubercle bacillus in a large number of picoliter droplets combined with a fluorescent probe has been developed. We use BlaC (an enzyme naturally expressed/secreted by tubercle bacilli) as a marker and a designed BlaC-specific fluorogenic substrates as probes for Mtb detection. We present here a new method to detect the indicator enzyme based on the isolation, digitization and concentration of bacteria samples in a large number of picoliter drops. We show that by controlling the size of the droplets we can control the rate of conversion. Hence rapid increase in signal has been observed as the size of the drops has been decreased.

Our vision is that this tool will be able to detect tubercle bacilli in a sensitive, rapid, specific and quantitative manner in vitro at a low cost, particularly in resource limited settings where TB is the most prevalent.

(627c) Microengineered, Photodegradable Hydrogels for the Selective Capture and Release of Mammalian Cells

Tibbitt, M. W., University of Colorado
Fischer, P., University of Wyoming
Kloxin, A. M., University of Delaware
Oakey, J., University of Wyoming
Anseth, K. S., University of Colorado-Boulder
Originally presented on: 11/7/2013 9:00:00 - 9:15:00

The capture and characterization of rare circulating tumor cells (CTCs) presents a unique opportunity in the development of personalized medicine and cancer treatment. Recently, microfluidic devices have been presented that sense and capture CTCs from complex biologic fluids, such as whole blood. Initially, high surface area post arrays were introduced,[1] followed by the design and implementation of a microvortex-generating herringbone channel with improved capture efficiency.[2] Currently, these devices are being applied clinically to diagnose and monitor cancer progression.[3] However, a significant limitation of CTC capture technologies is the inability to recover selectively captured cells in a viable state for downstream analysis. A few approaches have been proposed to mitigate this limitation, such as reversible alginate capture coatings[4] or thermally responsive capture surfaces.[5] These approaches, however, cannot be applied to the complex geometries needed for efficient cell capture and do not enable selective, individual cell release for on-chip purification and downstream analysis. Here, we present a photopolymerized, poly(ethylene glycol) (PEG) based hydrogel that can be photodegraded post-fabrication for the selective capture and release of cells from complex microfluidic geometries. To fabricate the microfluidic capture surfaces, a ~25 μm tall polydimethylsiloxane (PDMS) channel (straight or herringbone) was filled with a gel-forming monomer solution containing photodegradable PEG diacrylate (PEGdiPDA),[6] which was then photopolymerized using visible light. After the gel was formed, the initial PDMS mold was removed and the gel was capped with a ~75 μm tall PDMS straight channel. Then, the PEG gel was functionalized with capture antibodies. A suspension of A549 human lung carcinoma cells were flowed through the microfluidic channel and captured on the surface. Herringbone channels captured cells from media with a higher efficiency than straight channels. In each case, cells were released after capture by irradiated the gel with UV light under flow. The UV light degraded the gel completely and liberated the cells from the capture device so that they could be recovered downstream. In this manner, the PEG capture devices are employed for improved capture and purity of mammalian cells and may provide an avenue for earlier cancer detection as well as better diagnostics and prognostics.

References

1. Nagrath S *et al. Nature* **450** (2007) 1235-1239
2. Stott SL *et al. PNAS* **107** (2010) 18392-18397
3. Stott SL *et al. Sci Tran Med* **2** (2010) 25ra23
4. Shah AM *et al. Anal Chem* **84** (2012) 3682-3688
5. Hou S *et al. Adv Matls* **25** (2013) 1547-1551
6. Kloxin AM *et al. Nat Prot* **5** (2010) 1867-1887

(627d) Bihydrogel Particles As Free-Standing Mechanical Microsensors

Tsou, T. Y., National Taiwan University
Hsu, C. W., Chemical Engineering Department
Chen, H. Y., National Taiwan University
Hsieh, C. C., National Taiwan University
Originally presented on: 11/7/2013 9:15:00 - 9:30:00

We propose a new type of mechanical microsensors that are microparticles consisting of two parallel phases of crosslinked hydrogels with different sensitivity to a certain enviromental parameter. Working similarly to a bimetallic strip, this type of sensors can bend to a different degree depending on the corresponding environmental parameter and therefore was named bihydrogel microsensors. For demonstration, we have manufactured pH and temperature bihydrogel sensors that use acrylic acid and PNIPAM to form the pH and the temperature sensitive hydrogels. These microsensors were manufactured very efficiently and economically using the stop-flow lithography printing technique in PDMS microchannel. Because the bihydrogel microsensors are free-standing particles, they are particularly suitable for use in confined environment where most types of microsensors are unable to fit in. Furthermore, since the working principle of the bihydrogel microsensors is very general, the same principle can be applied for producing sensors for detecting a wide variety of environmental parameters.

(627e) Relative Humidity Optical Microsensor Based On Polymer Coated Microtoroids

Mehrabani, S., University of Southern California
Kwong, P., University of Southern California
Gupta, M., University of Southern California
Armani, A. M., University of Southern California
Originally presented on: 11/7/2013 9:30:00 - 9:45:00

Monitoring the relative humidity (RH) in the environment plays a major role in various applications ranging from fundamental science and industry to biomedical applications. For instance, detection, control and elimination of traces amounts of water vapor in moisture-sensitive environments such as semiconductor fabrication plants can increase the production yield significantly. Thus far, different optical and electronic techniques have been implemented for the relative humidity detection. However, electronic methods are sensitive to the ambient electromagnetic sources of noise therefore optical based techniques are more robust.

One of the optical detection techniques that has shown promising results in relative humidity detection is based on whispering gallery mode optical microresonators. In whispering gallery mode microresonators, light propagates inside a circular path confined at a specific resonant wavelength at the periphery of the device. Because the resonant wavelength is dependent on both the refractive index and the geometrical properties of the device, any changes to either of these characteristics can be detected by measuring the resonant wavelength shift. However, conventionally, the detection mechanism is based solely on a refractive index change.

In this study, we have developed an optical humidity sensing technique based on the combination of a whispering gallery mode silica microtoroidal cavity sensor and a novel nanometer thick humidity-responsive polymer coating, poly N-isopropylacrylamide (pNIPAAm). Once pNIPAAM is exposed to humidity, it induces a change in both the refractive index and the diameter of the hybrid microresonator, and therefore it is capable of improving the detection signal significantly. As a result of the conformational change by the polymer in response to humidity, the detection response of the mircoresonator was increased by nearly two orders of magnitude. Additional measurements exploring the hysteresis behavior of the sensing device, the response rate, the recovery time, and the effect of the pNIPAAm film thickness were also studied.

(627f) Sensitive Chemical and Biological Sensing Enabled By Self-Assembled Plasmonic Crystals

Gozubenli, N.,
Jiang, P., University of Florida
Originally presented on: 11/7/2013 9:45:00 - 10:00:00

We report the achievement of rapid and sensitive detection of various chemical and biological species including small molecules (e.g., benzenethiol), proteins, viruses and bacteria by using plasmonic crystals created by a simple and scalable bottom-up nanofabrication platform. This platform combines the simplicity and cost benefits of bottom-up colloidal self-assembly with the scalability and compatibility of top-down microfabrication for creating a large variety of highly ordered plasmonic nanostructures, such as periodic arrays of nanoholes, nanovoids, nanopyramids, and nanodimples. We have demonstrated the capability of this novel methodology in creating highly sensitive and reproducible surface-enhanced Raman scattering (SERS) and surface plasmon resonance (SPR) chemical and biological sensors over wafer-sized areas. The highest SERS enhancement factor and the SPR sensitivity achieved is on the order of 10^{10} and 800 nm per refractive index unit, respectively. This novel nanofabrication platform is compatible with standard microfabrication, enabling on-wafer integration of plasmonic biosensors with conventional microfluidic devices.

(627g) Multi Spatiotemporal Spectroscopy From a Novel Nano-Structured Plasmonic Grating Hotspots Platform

Basuray, S., University of Missouri
Pathak, A., University of Missouri
Mathai, J. C., University of Missouri
Menke, D., University of Missouri
Cornish, P. V., University of Missouri
Gangopadhyay, K., 3Nanos Technologies LLC
Gangopadhyay, S., University of Missouri
Originally presented on: 11/7/2013 10:00:00 - 10:15:00

Physiologically relevant processes to pharmacogenomics need to be simultaneously scrutinized, visualized and simulated to study local effects and molecular conformations for multiple applications from disease prognosis drug delivery to protein folding. However, a single biophysical instrument with sufficient spatial and temporal resolution to detect complex biological kinetics over broad spatiotemporal scales at the single molecule (SM) level is non-existent. In my talk today, I will show you a cost effective soft lithography technique to fabricate plasmonic structures that can lead to the development of multi spatial/temporal techniques from a single platform. The substrate are made using a simple PDMS stamping of grating structures from commercially produced HD-DVDs with a thermally deposited metallic cap. By modifying the metallic capping process using oblique (non-normal) angular thermal deposition, extra-ordinary (173.12 times) fluorescence enhancements is observed from the substrate than quartz for the same dye. A nano-patterned substrate with three major surface morphological features is obtained at definite deposition angles for a fixed deposition thickness. The surface consists of a shallow nano-gap region, a conical grating peak with a plateau region between the two aforementioned regions that leads to a concentration of the electric field due to antenna (nano-gap) and lightning rod (grating peak) effects. On closer inspection with SEM, AFM images show distinct secondary peaks on the grating peak that gives rise to an extreme concentration of the electric field, leading to localized surface plasmon resonance, called "hot-spots". This causes any fluorophore localized on a "hot-spot" to fluoresce with intensities that show higher enhancement in comparison to the bulk of the grating structure from the same substrate. We found that these substrates (specifically the nano-peaks) are ideally suited to study SM phenomenon like SM Förster resonance energy transfer (smFRET), due to an increased signal to noise ratio and high enhancement factors using a simple epi-fluorescence microscope. In addition, dyes close to the hotspot region show an increase in the Förster distance as a result of the localized SPR. The same plasmonic substrates are used also for Surface Enhanced Raman Spectroscopy (SERS). The Raman spectrum from a 1mM 2-naphthalenethiol self-assembled monolayer on the nano-grating structure at multiple spatial positions shows consistent signals with significantly enhanced Raman peaks. The spatial homogeneity (less than 10 % standard deviation) in signal enhancement of Raman Peaks clearly demonstrates the uniformity of the patterned nano-structure. In comparison with commercial substrates, we clearly observe higher enhancements of our substrate with regular spatial distribution of hot-spots with minimal signal variation in contrast to commercial substrate Klarite™ (high signal variation, no spatial uniformity, lower Raman signal strength). The enhancement was found to critically depend on the structural moiety and on the incident laser angle with correct optimization leading to SM-SERS. To conclude, we show here an inexpensive, easy method to obtain both uniform Raman hotspots that show large enhancement and extraordinary fluorescence enhancement with single molecule resolution. Combined with microfluidic architecture, an opto-fluidic device can be easily implemented to increase detection sensitivity and selectivity as a novel bio-diagnostic platform or to simultaneously observe pertinent and biological motions and kinetics from a single platform.

(627h) A Nove Microfluidic Technique for Detecting Clinical Isolates of Inluenza A Virus

Wang, J., Brown University
Tai, W.,
Tripathi, A., Brown University

Originally presented on: 11/7/2013 10:15:00 - 10:30:00

This work presents the clinical application of a robust and unique approach for RNA amplification, called a Simple Method for Amplifying RNA Targets (SMART), for the detection and subtype identification of H1N1 swine, H1N1 seasonal, and H3N2 seasonal influenza virus. While all the exciting amplification techniques rely on the diffusion of two molecules to complex RNA structures, the SMART method achieves fast and efficient amplification via single molecule diffusion. SMART utilizes amplifiable single stranded DNA (ssDNA) probes serve as reporter molecules for specific viral RNA (vRNA) sequences that are captured and separated on a microfluidic chip under zero-flow conditions. The assay is performed using an isothermal (41°C) amplification scheme via a modified version of nucleic acid sequence based amplification (NASBA). In our study, 116 consecutive, de-identified, clinical nasopharyngeal swab samples were analyzed independently and in a blinded fashion using SMART, PCR, antigen (Ag) testing, and quantitative viral culture. SMART correctly detected influenza in 98.3% of the samples with a subtyping accuracy of 95.7% for viral loads as low as 10^5 copies/mL and provides subtyping information unmatched to any other diagnostic test, while also performing at 60 times faster than the gold standard of viral culture. This work demonstrates that SMART represents a highly accurate, point-of-care (POC) diagnostic platform for the detection and subtyping of influenza virus in clinical specimens and offers significant advantages over current commercially available diagnostic tools.

(627i) Multiplexed Proteomics Using Ultra Dielectrophoresis (uDEP)

Emaminejad, S., Stanford University
Javanmard, M., Stanford Genome Technology Center
Dutton, R.,
Davis, R., Stanford Genome Technology Center
Originally presented on: 11/7/2013 10:30:00 - 10:45:00

Multiplexed Proteomics using Ultra Dielectrophoresis (uDEP)

Our goal is to develop a low cost electronic platform for multiplexed detection of protein biomarkers in a complex sample. Our platform is based on performing a bead based immunoassay, where along a single channel an array of antibodies is patterned. Below each element of the array is a pair of adressable interdigitated electrodes, which can detach the immunobound beads through negative dielectrophoresis (nDEP) force. The beads are detached region by region and then transported downstream where they are quantified electrically or optically. The main challenge with this technique lies in providing a strong enough force to detach the beads. Typically, nDEP provides on the order of a few picoNewtons of force, where the binding force between antibodies and antigens is on the order of hundreds of picoNewtons. By increasing the strength of the nDEP force, we demonstrated enhanced electrokinetic actuation that can be used to elute specifically-bound beads from the surface. When applying high voltages at the electrodes (> 10 V) that are in direct contact with the buffer, DEP force magnitude is limited by electrode corrosion due to electrochemical reactions at the interface of the electrodes and the solution. Using Atomic Layer Deposition we deposited a pinhole free nanometer-scale thin film oxide as a protective layer to prevent electrodes from corrosion. By exciting the electrodes at high frequency, we capacitively coupled the electrodes to the buffer in order to avoid electric field degradation, and hence, reduction in nDEP force due to the presence of the insulating oxide layer. Deposition of thin film oxide layer on the electrodes imposes a number of challenges. First challenge is degradation of the electric field and hence dielectrophoresis force, as a result of the undesired voltage drop across the oxide. To compensate for the voltage drop across the oxide, one may increase the applied voltage at the electrodes, but that may lead to breakdown of the oxide. By analytical derivation and as confirmed through characterization results, we showed that at sufficiently high frequencies (for our device > 1 MHz), the electric field across the oxide layer becomes independent of the thickness, and hence oxide breakdown does not impose limitation on the thickness of the film. Our fabricated electrodes are able to withstand voltages up to 120 Vpp, beyond which bubble formation inside the channel becomes the limiting factor. This results in two orders of magnitude improvement in DEP force, than what was possible with bare gold electrodes. Using this ultra-DEP (uDEP) device, we demonstrated 100% detachment of anti-IgG and IgG bound beads. The enhanced switching performance shows orders of magnitude of improvement in on-to-off ratio and switching response time, without need for chemical eluting agents, as compared to previous work. In this talk, we will present the operation of the uDEP-based device to perform a multiplexed assay.

(627j) Simultaneous Measurement of pH and Bacterial Markers in a Single Sensor

Sismaet, H. J., Northeastern University
Webster, T. A., Northeastern University
Goluch, E. D., Northeastern University
Originally presented on: 11/7/2013 10:45:00 - 11:00:00

Pseudomonas aeruginosa is one of the leading causes of gram-negative bacterial infection in hospitals. Significantly impacting people with compromised immune systems, *Pseudomonas aeruginosa* is an important bacterium to study both in the medical and scientific communities. Unique to this species is its production of pyocyanin, a redox-active quorum sensing molecule (QSM) linked to biofilm formation. Because it is redox-active, pyocyanin can be directly detected electrochemically in solution. We have shown that it possible to detect this QSM with very high sensitivity using microfabricated nanofluidic electrode assemblies (NEAs). The development of these devices has ultimately led to a growing interest in their medical applications. One potential application is the integration of these sensors in smart bandages, a type of wound dressing that will incorporate an array of sensors for early diagnosis of infections. The reference electrode integrated in our NEAs is uniquely sensitive to both proton concentration and electro-active species, making it possible to simultaneously monitor pH and pyocyanin concentration using a single sensor.

To test the device's dual ability to monitor pH and pyocyanin, samples of 100 ÂµM pyocyanin in 100 mM phosphate buffer were created at different pH values (4.3, 7.0, and 10.4) through the addition of HCl acid or NaOH base. Samples were loaded into a NEA connected to a potentiostat (CHI842C) to control the applied potentials, which ranged from -0.75 to 0.00 V, vs. the integrated palladium hydride reference electrode. Using square-wave voltammetry to detect the presence of pyocyanin, all sample scans were performed at an amplitude voltage of 0.05 volts and a frequency of 15 Hz. As the pH of the sample was varied, a shift in the maximum peak potential was observed. Further analysis showed a linear correlation between pH and the peak potential with a change of 37 mV per pH unit. The ability to simultaneously measure the concentration of pyocyanin and the pH of a fluid sample highlights the device's future medical application as part of a smart bandage where pH is linked to general bacterial infection and the presence of pyocyanin indicates the existence of *Pseudomonas aeruginosa*. Other electro-active molecules can also be detected by changing the applied potentials and functionalizing the sensor to potentially detect other bacterial species as well.

AUTHOR INDEX

AUTHOR INDEX

American Institute of Chemical Engineers
3 Park Avenue
New York, New York 10016-5991

CURRAN ASSOCIATES INC
proceedings.com